Published by J. B. LIPPINCOTT COMPANY, PHILADELPHIA·NEW YORK

Printed in the United States of America
Library of Congress Catalog Card Number 26-18089

Contents

PART ONE

PART TWO

PART I

THE FIRST CHAPTER

The Animal Shop

THIS book of the memoirs of Doctor Dolittle has been called the *Caravan* because it is in part a continuation of the Circus and the adventures that he met with in his career as a showman. Moreover, on his arrival in London the headquarters of the Dolittle household became the Doctor's caravan on Greenheath (just outside the city) where his surgery and animal clinic continued their good work. And this too made that name for the book seem proper and in place.

It will be remembered that shortly after John Dolittle was elected as the new manager by the performers in Blossom's old circus he had received a special invitation from some theater owners in London to come to that city and put on a show for them. And while he was still fulfilling various engagements in small towns and working (after Blossom's scandalous departure with all the money that had been earned in Manchester) to get enough cash in hand to put the circus on its feet again, he was continually trying to think out in spare moments some good and original show to put on in London.

He was most anxious that his company's first appearance in the big city should be a success. The staff of the Dolittle

Circus now consisted of Matthew Mugg, Assistant Manager; Hercules the Strong Man; The Pinto Brothers, trapeze artists; Hop the Clown; Henry Crockett, the Punch-and-Judy man; Theodosia Mugg, Mistress of the Wardrobes; and Fred, a new menagerie keeper whom the Doctor had lately engaged. Then of course there were the animals: the lion, the leopard and the elephant, the big animals that constituted the important part of the menagerie; several smaller beasts, like the opossum (whom Blossom had styled the "Hurri-gurri"); the Pushmi-Pullyu; the snakes; the Doctor's own animal household (Jip the dog, Gub-Gub the pig, Too-Too the owl, Dab-Dab the duck, and the white mouse); and a few other oddments.

It was not a large company. And for this John Dolittle was grateful. In the hard days that followed Blossom's heartless desertion it was difficult enough to earn sufficient money to buy food for even these. But they were all, without exception, extremely sportsmanlike when things went badly. On the Doctor's new plan (the "cooperative system" as he called it) all the staff shared in the profits instead of getting wages. Often when business was poor this had meant no salary at all, nothing but three meals a day. Yet there were no desertions and no grumbling. Every one felt sure that sooner or later John Dolittle would steer the company's ship into prosperous waters, and they stuck to him through thick and thin. And the day came when their confidence was justified.

The way the Doctor finally hit upon an idea for a novel show for his London engagement was rather curious. Like many important things, it began from a small chance happening.

One evening, when the show had moved to a moderate-sized market town, the Doctor went for a walk with Matthew

Mugg and Jip. They had been busy all day getting
set up and the Doctor had not yet had an oppor
see anything. After going through the main streets, t
to an inn which had tables and chairs set outside be... the

"They sat down to drink a glass of ale"

door. It was a warm evening and the Doctor and Matthew sat down at the inn tables to drink a glass of ale.

While they were resting and watching the quiet life of the town the song of a bird reached their ears. It was extraordinarily beautiful, at times tremendously powerful, at others soft and low and mysterious—but always changing. The singer, whoever he was, never repeated himself.

The Doctor had written books on bird songs and he was interested.

"Do you hear that, Matthew?" he asked.

"Great, ain't it?" said the Cats'-Meat-Man. "Must be a nightingale—up on them big elms by the church there."

"No," said the Doctor, "that's no nightingale. That's a canary. He is singing scraps of a nightingale's song which he has picked up—and parts of many others, too. But he has a canary's voice, for all that. Listen; now he's imitating a thrush."

They sat a while longer and the bird ran through a wonderful range of imitations.

"You know, Matthew," said the Doctor, "I think I'd like to have a canary in the wagon. They're awfully good company. I've never bought one, because I hate to see birds in cages. But with those who are born in captivity I suppose it's really all right. Let's go down the street and see if we can get a glimpse of this songster."

So after the Doctor had paid for the ale they left the tables and walked along toward the church. But before they reached it they saw there were several shops to pass. Presently the Doctor stopped.

"Look, Matthew," said he. "One of those shops is an animal shop. That's where the canary is. I hate animal shops; the poor creatures usually look so neglected. The proprietors always keep too many—more than they can look after properly. And they usually smell so stuffy and close—the shops, I mean. I never go into them now. I don't even pass one if I can help it."

"Why?" asked Matthew.

"Well," said the Doctor, "ever since I became sort of known among the animals the poor beasts all talk to me as soon as I go in, begging me to buy them—birds and rabbits and guinea pigs and everything. I think I'll turn back and go around another way, so I won't have to pass the window."

But just as the Doctor was about to return toward the inn the beautiful voice of the song bird burst out again and he hesitated.

"He's marvelous," said John Dolittle—"simply superb!"

"Why not hurry by with just one eye open?" said Matthew. "Maybe you could spot the bird without stopping."

"All right," said the Doctor. And putting on a brisk pace, he strode toward the shop. In passing it he just gave one glance in at the window and hurried on.

"Well," asked Matthew, as the Doctor paused on the other side, "did you see which bird it was?"

"Yes," said John Dolittle. "It's that green canary near the door, the one in the small wooden cage, marked three shillings. Listen, Matthew, go in and buy him for me. I can afford that much, I think. I dare not go myself. Everything in the place will clamor at me at once. I have an idea those white rabbits recognized me already. You go for me. . . . Don't forget—the green canary in the wooden cage near the door, marked three shillings. Here's the money."

So Matthew Mugg went into the store with the three shillings, while the Doctor waited outside the window of the shop next door.

The Cats'-Meat-Man wasn't gone very long—and when he returned he had no canary with him.

"You made a mistake, Doctor," said he. "The bird you spoke of is a hen and they don't sing. The one we heard is a bright yellow cock, right outside the shop. They want two pounds ten for him. He's a prize bird, they say, and the best singer they ever had."

"How extraordinary!" said the Doctor. "Are you sure?"

"The Doctor waited outside the window of the shop next door"

And, forgetting for the moment all about his intention of not being seen by the animals in the shop, he moved up to the window and pointed again to the green canary.

"That's the bird I meant," he said. "Did you ask about

that one?—Oh, Lord! Now I've done it. She has recognized me."

The green canary near the door-end of the window, seeing the famous Doctor pointing to her, evidently expected him to buy her. She was already making signs to him through the glass and jumping about her cage with joy.

The Doctor, quite unable to afford two pounds ten for the other bird, was beginning to move away. But the expression in the little green canary's face as she realized he didn't mean to buy her after all was pitiful to see.

John Dolittle had not walked with Matthew more than a hundred yards down the street before he stopped again.

"It's no use," said he. "I'll have to buy her, I suppose—even if she can't sing. That's always the way if I go near an animal shop. I always have to buy the wretchedest and most useless thing they have there. Go back and get her."

Once more the Cats'-Meat-Man went into the shop and presently returned with a small cage covered over with brown paper.

"We must hurry, Matthew," said the Doctor. "It's nearly time that the tea was served and Theodosia always finds it hard to attend to it without our help."

On reaching the circus the Doctor was immediately called away on important business connected with the show. He asked Matthew to take the canary to the wagon, and he was himself occupied with one thing and another until supper time.

And even when he finally returned to his wagon his mind was so taken up with the things of the day that he had forgotten for the moment all about the canary he had bought. He sank wearily into a chair as he entered and Too-Too, the accountant, immediately engaged him in a financial conversation.

But the dull discussion of money and figures had hardly begun before the Doctor's attention was distracted by a very agreeable sound. It was the voice of a bird warbling ever so softly.

"Great heavens!" the Doctor whispered. "Where's that coming from?"

The sound grew and grew—the most beautiful singing that John Dolittle had ever heard, even superior to that which he had listened to outside the inn. To ordinary ears it would have been wonderful enough, but to the Doctor, who understood canary language and could follow the words of the song being sung, it was an experience to be remembered.

It was a long poem, telling of many things—of many lands and many loves, of little adventures and great adventures, and the melody, now sad, now gay—now fierce, now soft, was more wonderful than the finest nightingale singing at his best.

"Where is it coming from?" the Doctor repeated, completely mystified.

"From that covered cage up on the shelf," said Too-Too.

"Great heavens!" the Doctor cried. "The bird I bought this afternoon!"

He sprang up and tore the wrapping paper aside. The song ceased. The little green canary peered out at him through the torn hole.

"I thought you were a hen," said the Doctor.

"So I am," said the bird.

"But you sing!"

"Well, why not?"

"But hen canaries don't sing."

The little green bird laughed a long, trilling condescending sort of laugh.

"That old story—it's so amusing!" she said. "It was invented by the cocks, you know—the conceited males. The hens have by far the better voices. But the cocks don't like us to sing. They peck us if we do. Some years ago a move-

"She laughed a long trilling laugh"

ment was started—'Singing for Women,' it was called. Some of us hens got together to assert our rights. But there were an awful lot of old-fashioned ones—old maids, you know—who still thought it was unmaidenly to sing. They said that a hen's place was on the nest—that singing was for men only. So the movement failed. That's why people still believe that hens *can't* sing."

"But you didn't sing in the shop?" said the Doctor.

"Neither would you—in *that* shop," said the canary. "The smell of the place was enough to choke you."

"Well, why did you sing now?"

"Because I realized, after the man you sent came in a second time, that you had wanted to buy that stupid yellow cock who had been bawling out of tune all afternoon. I knew, of course, that you only sent the man back to get me out of kindness. So I thought I'd like to repay you by showing you what we women can do in the musical line."

"Marvelous!" said the Doctor. "You certainly make that other fellow sound like a second-rate singer. You are a contralto, I see."

"A mezzo-contralto," the canary corrected. "But I can go right up through the highest soprano range when I want to."

"What is your name?" asked the Doctor.

"Pippinella," the bird replied.

"What was that you were singing just now?"

"I was singing you the story of my life."

"But it was in verse."

"Yes, I made it into poetry—just to amuse myself. We cage-birds have a lot of spare time on our hands, when there are no eggs to sit on or young ones to feed."

"Humph!" said the Doctor. "You are a great artist—a poet and a singer."

"And a musician!" said the canary quietly. "The composition is entirely my own. You noticed I used none of the ordinary bird songs—except the love song of the greenfinch at the part where I am telling of my faithless husband running off to America and leaving me weeping by the shore."

Dab-Dab at this moment came in to announce that supper was ready, but to Gub-Gub's disgust the Doctor brushed everything aside in the excitement of a new interest. Diving into an old portfolio, he brought out a blank musical manuscript book in which he sometimes wrote down pieces for the flute, his own favorite instrument.

"Excuse me," he said to the canary, "but would you mind starting the story of your life all over again. It interests me immensely."

"Certainly," said the little bird. "Have my drinking trough filled with water, will you please? It got emptied with the shaking coming here. I like to moisten my throat occasionally when I am singing long songs."

"Of course, of course!" said the Doctor, falling over Gub-Gub in his haste to provide the singer with what she wanted. "There! Now, would you mind singing very slowly? Because I want to take down the musical notation and the time is a little complicated. Also, I notice you change the key quite often. The words I won't bother with for the present, because I couldn't write both at once. I will ask you to give me them again, if you will, later. All right. I'm ready whenever you are."

THE SECOND CHAPTER

The White Persian

THEN the Doctor sat down and wrote page after page of music while the green canary sang him the story of her life. It was a long song, lasting at least half an hour. And during the course of it Gub-Gub interrupted more than once with his pathetic—

"But, Doctor, the supper's getting cold!"

When she had finished John Dolittle carefully put away the book he had been writing in, thanked the canary and prepared to have supper.

"Would you care to come out of your cage and join us?" he asked.

"Have you any cats?"

"No," said the Doctor. "I don't keep any cats in the wagon."

"Oh, all right," said the canary, "then if you'll open my door I'll come out."

"But you could easily get away from a cat, couldn't you," asked Jip, "with wings to fly?"

"I could if I was expecting it or knew where it was," said the canary, flying down onto the table and picking up a crumb beside the Doctor's plate. "Cats are most dangerous when you can't see them. They are the only really skilful hunters."

"Huh!" grunted Jip. "Dogs are pretty good, you know."

"Excuse me," said the canary, "but dogs are mere duffers when compared with cats in the hunting game—I'm sorry to hurt your feelings, but *duffers* is the only word I can use. You are all very fine at following and tracking—even better than cats at that. But for getting your quarry by the use of your wits—well, there! Did you ever see a dog sit and watch a hole in the ground for hours and hours on end, silent and still as a stone—waiting, waiting for some wretched little mouse or other creature to come out? Did you ever know a dog with the patience to do that? No: your dog, when he finds a hole, barks and yelps and scratches at it—and of course the rat, or whatever it is, doesn't dream of coming out. No, speaking as a bird, I'd sooner be shut in a roomful of dogs than have a single cat in the house."

"Did you ever have any unpleasant experience with them?" asked the Doctor.

"Myself, no," said the canary. "But that was solely because of someone else's experience with one. It taught me a lesson. I lived once in the same house with a parrot. One day the woman who owned us got a fine, silky, white Persian. She was a lovely creature—to look at. Said the old parrot to me the morning the cat came: 'She looks a decent sort.'

" 'Pol,' said I, 'cats are cats. Don't trust her—never trust a cat.' "

"I wonder if that's what makes them the way they are," said the Doctor—"the fact that no one ever trusts them. It's a terrible strain on anybody's character."

"Fiddlesticks!" said the canary. "Our woman trusted this cat—even left her in the room with us at night. My cage was hung high up on a chain, so I wasn't afraid of her reaching in to me with her claws. But poor old Pol, one of the decentest old cronies that ever sat on a perch, he had no cage at all—just one of those fool stands they make for parrots—a cross-bar perch and a long chain on his ankle. He wouldn't believe that this sweet creature in white was dangerous, until one day she tried to climb up the pole of his stand and get at him. Well, a parrot's a pretty good fighter when the fight's a fair one, and he gave her more than she

"He gave her more than she bargained for"

bargained for. She retired from the fray with a piece bitten out of her ear.

" 'Now, will you believe me?' I said. 'And, listen; she's going to get you yet—if there's any way to do it, that she and the Devil can think up between them. Whatever you do, don't go to sleep while she is in the room. She's scared of you now, while you're facing her. But she won't be scared of you as soon as you're off your guard. One spring and a bite on the neck from her and Polly won't want any more crackers. Remember—*don't go to sleep when she is in the room.*'"

The green canary paused a moment in her story to hop across the table and take a drink out of Gub-Gub's milk bowl—which piece of cheek greatly astonished that member of the household. Then she cleaned her bill against the cruet stand and proceeded:

"I couldn't tell you how many times I saved that foolish parrot's life. An easy-going bird, he loved regularity. He was a bachelor, making a great ceremony of all the little habits of his daily round. And he just couldn't bear to have anything interfere with them. He would be ruffled and sulky for days if the maid missed giving him his bath on Saturday afternoon or his piece of orange peel at Sunday breakfast. One of his little customs was to take a nap every day after lunch. I warned him over and over again that this was dangerous unless the doors and windows were shut and the cat outside. But the force of habit, years and years of bachelor regularity, were too strong for him. And I believe he would have taken that nap if the room had been full of cats."

The canary picked up another crumb, munched it thoughtfully and went on:

"I often think there was something fine about that parrot's independence. He had principles and nothing was allowed to change them. In the meantime that horrible cat was waiting for her chance. Often and often when Pol was dozing off I'd see her come sneaking toward his stand along the floor or creep across a table near enough to spring from. Then I'd give a terrific, loud whistle and the parrot would wake up. And the cat would slink away, looking daggers at me for spoiling her game.

"She took a drink out of Gub-Gub's milk bowl!"

"As for the mistress we had, it never entered her empty head that the cat was a dangerous customer. One day a friend of hers asked if she wasn't afraid to leave the beast around when she had no cage over the parrot.

" 'Oh, tut, tut!' said she. 'Pussums wouldn't hurt my nice Polly, would ums, Pussums?'

"And then that silky hypocrite would rub her neck against the old lady's dress and purr as though butter wouldn't melt in her mouth.

"Well, I did my best. But the day came when even I was outwitted by the she-devil in white. The old lady had gone to visit friends in the country and let the maid take the day off while she was away. Both the parrot and I were given double rations of seed and water, the house was locked and the keys put under the mat. The door of the parlor, where we were always kept, was closed, and I thanked my stars that for this day, anyhow, my friend should be safe.

"About noon a thunder storm came up and the wind howled around the house dismally. And presently I saw the door of our room blow open. It had not been properly latched—just closed carelessly.

" 'Don't go to sleep, Pol,' I said. 'That cat may come in any moment!'

"Well, for a long time she didn't. And after an hour I decided that the cat must have been shut in another room somewhere and that it was all right and I needn't worry. After his lunch Pol went sound asleep; and presently, feeling sort of drowsy myself, I too took a nap.

"I dreamed all sorts of awful things—monstrous cats leaping through the air, parrots defending themselves with swords and pitchforks—all manner of terrible stuff. At the most tragic moment in the worst dream I thought I heard a thud on the floor and suddenly woke up, wide awake, the way one does with nightmares. And there on the floor lay Pol, stone dead, and squatting on the carpet on the far side of him, staring up at me with a devilish smirk of glee on her horrible face, sat the white cat!"

The canary shivered a little and rubbed her bill with her right foot, as though to wipe away the memory of a bad dream.

"I was too horrified to say a word," she presently continued, "and I began to wonder whether the abominable wretch would eat my poor dead friend. But not a bit of it. She didn't want him for food at all. She got three square meals a day from the old lady, the daintiest morsels in the house. She just wanted to kill—to kill for the fun of killing. For three months she had watched and waited and calculated. And in the end she had won. With another grin of triumph in my direction, she slowly turned about, left the body where it lay and stalked toward the door.

" 'Well,' I thought to myself, 'there's one thing: she can't escape the blame. At least the old lady will know her now for what she is, the murderess!'

"And then a curious thing happened. It reminded me of something my mother used to believe: that cats are helped by the Devil. 'It would be impossible for them to be so fiendishly clever without,' she used to say. 'Never try to match your wits against a cat! They are helped by the Devil.'

"I had never believed it, myself. But that afternoon I came very near believing it. Now, mark you, with that door blown open, any one would know that it was the cat who had come in and killed the parrot, wouldn't they? But with that door shut—the way the old lady had thought she left it—and the cat outside of the room—no one could possibly suspect 'sweet pussums.' So then I felt quite certain that this time the cat was going to get in a good, stiff row. Now comes the queer business; no sooner had she passed into the hall outside than the wind began again, howling and moaning about the house. And, to my horror, I saw the door slowly closing. Faster and faster it swung forward, and then with a bang that shook the house from cellar to garret, it slammed shut. The last glimpse I got of the hall outside showed me 'sweet pussums' squatting on the floor, still grinning at me in triumph. After that, I think you will admit, it was pardonable to believe that she

"The old lady gave her a saucer of milk"

was helped by the Devil. For, mind you, if the wind had come two minutes earlier it would have shut her *inside* the room, instead of *out*.

"Of course, when the old lady came home she just couldn't

understand it. There lay the parrot on the floor, his neck broken (the cat had done it very neatly and cleverly—just one spring, a bite and a twist); the windows were shut; the door was shut.

"Finally that stupid old woman said that perhaps boys had got in, probably down the chimney, wrung the parrot's neck and escaped, leaving no tracks. The mystery was never solved. She was frightfully upset, weeping all over the place—after it was too late.

" 'Oh, well,' she sobbed, 'anyhow, I have my canary left—and my sweet pussums.'

"And then that she-devil came up to her, purring, to be petted, and the old woman gave her a saucer of milk! No, never, never trust a cat."

"They're funny creatures," said the Doctor. "There's no gainsaying that. And their curious habit of killing even when they're not hungry is very hard to explain. Still, it's in their nature, I suppose, and one should never judge any one without making allowances for the nature he was born with. You have been through some very interesting experiences, I see. When you were singing me the story of your life I was so busy getting the music down that I couldn't pay much attention to the words. After we have finished supper, would you mind telling it to me over again?"

"Why, certainly," said the canary. "I'll tell it to you conversationally—without music."

"Yes, I think that would be better," said the Doctor. "You can then put in all your adventures in detail, without bothering to make the lines scan and rhyme. Gub-Gub, as soon as you have finished that plateful of beechnuts we will let Dab-Dab clear away."

THE THIRD CHAPTER

An Animal Biography

IT WAS thus through the coming of the little green canary that the Doctor wrote the first of his animal biographies. He had frequently considered doing this before. He claimed that in many instances the lives of animals were undoubtedly more interesting—if only they were properly written—than the lives of some of our so-called great men. He had even thought of writing a series, or a set, of books called *Great Animals of the Nineteenth Century,* or something like that. But so far he had not met many whose memories were good enough to remember all the things in their lives that make a biography interesting.

Gub-Gub, disappointed that no statue had been erected to him in Manchester, had often begged the Doctor to write his life for him, feeling certain that of course everybody would want to read it. But John Dolittle and his pets knew Gub-Gub's life by heart already. And, while the Doctor felt that it would make good comic reading, Gub-Gub himself refused to have it written that way, now that he was a famous actor.

"I want a dignified biography," said he. "I may be funny on the stage—very funny. But in my biography I must be dignified."

"Pignified, you mean," growled Jip. "Your biography would be just one large meal after another—with stomachaches for adventures. Myself, I'd sooner read the life of a nice, round, smooth stone."

So this branch of the Doctor's natural history writing had remained untouched till the appearance of Pippinella, the canary who came to join his family circle under such curious circumstances. Pippinella, the Doctor often said, was a born biographer, for she had a marvelous memory for the little things that made a story interesting and real. And John Dolittle in the preface to this the first of his *Private Memoirs of Distinguished Animals* was careful to say that the entire book was Pippinella's own, he merely having translated it from Canary into English.

Those who read it declared it most interesting. But, like so many of the Doctor's works, it is now out of print and copies of it are almost impossible to obtain. One of the reasons for this was that the ordinary booksellers wouldn't keep it. "Pooh!" they said. "*The Life of a Canary!* What kind of a life could that be—sitting in a cage all day?"

And as a consequence of their stupidity the book was only sold at the taxidermists' shops, naturalists' supply stores and odd places like that. Probably that is why copies are so hard to find to-day. In its final completed form, under the title of *The Life of Coloratura Pippinella, Contralto Canary,* the story contained much of the bird's life that was lived after she joined the Dolittle household. Moreover, the Doctor went through the manuscript with the authoress several times and got her to tell him more about many little incidents and details which he thought would be of interest to the general public. All this went to make it quite a long book. I have not space to set it down for you here just as the Doctor wrote it, but I will tell it you in part, at all events, as Pippinella herself related it to John Dolittle and his family circle.

"People," Pippinella began, when Gub-Gub had finally ceased fidgeting, "might think that the biography of a cage canary would be a very dull monotonous story. But, as a matter of fact, the lives of cage-birds are often far more varied and interesting than those of wild ones. I have heard the lives of several wild birds and they were mostly exceedingly dull and samey. Very well, then, I will begin at the beginning: I was born in an aviary—a private one, occupied by our family and a few others. My father was a bright lemon-yellow Harz Mountains canary and my mother was a greenfinch of very good family. My brothers and sisters—there were six of us altogether, three boys and three girls—were about the same as me to look at, sort of olive green and yellow mixed up. Of course, until our eyes were open the thing that concerned us chiefly was getting enough food. Good parents—and ours were the most conscientious couple you ever saw—give their children when they are first hatched about fourteen meals a day."

"Huh!" muttered Gub-Gub. "That's more than I ever got."

"Sh!" said the Doctor. "Don't interrupt."

"Pardon me, Pippinella," said John Dolittle, "but that is a point that has often interested me: How do young birds know, before their eyes are open, when their parents are bringing food? I've noticed that they all open their mouths every time the old birds come back to the nest."

"We tell it, I imagine, by the vibration. Our parents stepping on to the edge of the nest is something that we get to recognize very early. And then, although our eyes are closed, we see the shadow that our parents make leaning over the nest coming in between us and the sunlight."

"Thank you," said the Doctor, making a note. "Please continue."

"As you may have observed," Pippinella went on, "young birds talk and peep and chirp almost immediately they are out of the egg. That is one of the big differences between bird children and human children: you see before you talk, and we talk before we see."

"Huh!" Gub-Gub put in. "Your conversation can't have much sense to it then. What on earth can you have to talk about if you haven't seen anything yet?"

"That," said the canary, turning upon Gub-Gub with rather a haughty manner, "is perhaps another important difference between bird babies and pig babies: we are born with a certain amount of sense, while pigs, from what I have observed, never get very much, even when they are grown up. No, this blind period with small birds is a very important thing in their education and development. You ask me what they talk about. Nothing very much. I and my brothers and sisters used to swap guesses with one another on what the world would look like when our eyes would open. But the value of that time lies in this: having to do without our eyes, we develop what we call our sixth sense. It is rather hard to explain. But Too-Too will tell you that it is something well known and recognized among all birds. When we talk of birds having sense we always mean sixth sense."

"Excuse me," the Doctor put in, "but would you go into that a little further?"

"Certainly," said the canary. "But, as I told you, it's frightfully hard to explain. You were speaking just now of the parents approaching the nest and the young ones opening their mouths. Well, even before they actually step on the nest we soon get to know that they are there without seeing or hearing them. And then birds are awfully busy with their ears during this time. They do a lot of listening. And, being unable to see, they get to be much better hearers than if they had the use of their eyes as well. We listened to everything with the greatest care, trying to learn from it what the world was going to look like—even to the mice scratching behind the panelings and the boughs of the trees in the garden tapping the window-pane near our cage.

" 'That's a strong wind, isn't it, Father?' we would say. 'There are many twigs scratching on the glass.'

" 'Yes, children,' he would answer. 'It's a north wind. It is only the north and the northeast winds that press the jasmine up against the window. The others blow it away from the house.'

"Then, after that, you see, we could tell just by the tune the bough tips played upon the panes which way the wind was blowing and how strong it was. But much of our education we got without knowing the why or the wherefore at all. That is perhaps the best explanation of the sixth sense: just knowing a thing without knowing why or how you know it. Of course, in many matters the wild birds are much cleverer than we are. At geography, for instance: bird geography is all done by the sixth sense. But then, of course, we cage-birds don't get much chance to study that. But at other things we are ahead of them a long way. Especially about people. You'd be surprised what a good judge of character most housebred canaries are. Altogether as a grownup bird I've often astonished myself at what a lot I know—and how I know it I couldn't tell you to save my life. But I'm convinced that a great deal of the most important part of my education came to me, as it does to all of us, during that time when we lie in the nest with our eyes closed, trying to guess, by hearing and smelling and feeling, what the world is going to look like."

"Thank you," said the Doctor. "That is very interesting. Pray pardon my interruption. But these things are important to me as a naturalist. Please continue with your story."

"It is quite an exciting moment," Pippinella continued, "for young birds when they first open their eyes. They usually stay awake most of the night before, lest they sleep past the time and their brothers and sisters crow over them that they saw the world first.

"Well, with us the day came in due course; and, myself, I was slightly disappointed. You must remember that for us cage-birds the world was the inside of a room, instead of an open meadow, hedgerow or leafy forest. Of course, we had known something of what it was going to look like from asking our parents. But, no matter how well a thing is described to you, you always form your own idea of it, more or less wrong. Very well, then. Our world, we discovered, was a long room, sort of parlor and conservatory combined—a pleasant enough place, containing flower pots, palms, some furniture and several cages of birds. By day a woman attended to us, supplying our parents with chopped egg and cracker crumbs, which they fed to us youngsters. At night a man, who apparently owned us, came in to inspect everything. He seemed a decent sort of fellow and evidently had our welfare at heart, because he was forever rowing the woman for neglecting to clean cages, to change the water in the troughs or to give the birds fresh lettuce.

"Raising prize singing birds was this man's hobby. He had other cages in other rooms because we could hear the birds singing. And when the doors were open my father would sing back to them and carry on conversations with them about the woman, the quality of the new supply of seed, the temperature in the conservatory and any odd gossip about the household.

"There was another family of young ones in a cage close to ours. And our parents used to chat with the other parents —my mother always boasting that we looked a much healthier brood than theirs.

"The man had two children of his own, and they would

come into the conservatory occasionally to look at us and to play with toys on the floor. Their games provided us with entertainment and we were glad to have them, because most days the conservatory was rather quiet.

"My father was evidently quite a fine singer. And now and then, when he wasn't helping my mother shovel food into us hungry children, he'd sit on the edge of the nest and sing. He had a tremendous voice. But, for my part, I can't say that I enjoyed it much. At that close range it was simply ear-splitting and we used to beg Mother to make him stop.

"Once he was taken away from us for a whole day. And Mother told us he had gone to a canary show, to see if he was a good enough singer to get a prize. And when the man brought him back to us in the evening there was great excitement throughout the house. All the family came in, talking. Father had won the first prize at the show! After that he used to sing louder than ever, and it was no use our asking Mother to stop him, because she was even prouder of his success than he was. In between songs he told us all about the show and what the judges said and what sort of canaries he had had to sing against.

"It was a funny life—not nearly as dull as you'd think. As we got our feathers and grew bigger I used to look out of the window at the spring trees budding in the garden. Every once in a while I'd see a finch fly by and I'd get a sort of vague hankering to be out in the open, living a life of freedom. But one day I saw a hawk swoop down on a poor lame sparrow and carry him off. With a shudder I nestled down among my brothers and sisters and thanked my stars that I lived indoors. After all, I decided, there was a good deal to be said for this cage life, where we were protected from cats and birds of prey, given comfortable quarters and the best of food."

THE FOURTH CHAPTER

Pippinella Takes Her First Journey

"THE thing," Pippinella continued, "that most interested every bird born at the fancier's was whether he was going to be kept, sold, exchanged or given away. For this man, you must understand, while he did not run a regular shop, had many friends visit him who were, like himself, interested in canaries. He seems to have been quite well known, for some of these people came long distances to see him and his birds. His place wasn't big enough to keep all the families that were born there. So when the young ones grew up and got their full feathers he would pick out those that he wanted to keep himself. These were such as had good voices or who were prettily marked. The others he would sometimes exchange, sometimes give away and sometimes sell. He never kept very many hens.

"One evening about a fortnight after we had left the nest and were hopping and picking for ourselves my parents were talking this matter over together. And of course we youngsters, since it was a subject that very deeply concerned us, were listening intently. Said my mother:

" 'I'm afraid he will probably get rid of most of this brood. Nearly all his space is taken up and he seems to prefer those birds over in the next cage—though what he can see in the scrawny, long-necked little brats I don't know. I wouldn't exchange one of our babies for the whole batch.'

" 'Well,' said my father, 'so far as the welfare of our own is concerned, it will be just as well if he does let them go—especially if they go separately.'

" 'Why?' I asked.

" 'Because you always get better cared for in houses where you are the only canary kept. In any place where they have an awful lot to look after the treatment is usually slipshod and negligent. The worst of all are the animal shops. They are notoriously bad. You don't get your cage cleaned out more than once a week; you get put any place, sometimes in the hot sun, sometimes in an awful draught. And the noises and the smells are dreadful. No, I hope, for your sake, you don't get sent to an animal shop.'

" 'But, Father,' I said, 'it's all right if you get bought right away, isn't it?'

" 'Yes, but you seldom are, if you're a hen,' he said. 'People don't often come to an animal shop to buy hen canaries.'

" 'Why?' I asked again.

" 'Because they don't sing,' said he.

"You notice he said '*don't* sing,' not '*can't* sing.' I am afraid I've always been something of a rebel. Maybe I ought to have been born a cock. Anyway, that evening I felt particularly aggrieved at this stupid, unfair, old-fashioned custom.

" 'Father, I think that's ridiculous,' I said. 'You know very well that hens are born with just as good voices as cocks. But merely because it isn't considered proper for them to sing they have to let their voices spoil for want of practise when they're young. I think it's a crying shame.'

"Then my mother joined in.

" 'How dare you speak to your father like that, you brazen hussy!' she cried. 'What are the girls coming to these days, I'd like to know? Go and stand down in the corner of the cage!'

"And she gave me a box on the ear with her wing that knocked me right off the perch.

"Well, although I had been reprimanded, I was by no means repentant. I saw that just because hens were not supposed to sing I and my sisters stood a good chance of being packed off to some wretched crowded animal shop, instead of being bought by some private person who would treat us decently. And I determined to practise my singing secretly, so as to develop my voice and become just as valuable as my brothers.

"Well, in spite of frequent peckings, I continued to exercise my voice quietly when the others were busy eating or talking together. Finally the man who owned us noticed that I often got sat upon by the rest of my family and I was put in a separate cage. After that I could sing as long and as loud as I wanted to and all that the others could do was to make rude remarks from across the room about the quality of my voice.

"And then one day the fancier brought in a friend of his to see us. He wanted to make this friend a present of a canary, it seemed, and he offered him his choice out of the two new families of birds. I liked the man's face and I was determined he should pick me out, if I could make him. He was evidently rather taken with the coloring of the other family and he lingered around their cage quite a while. But I sang my loudest and my best and presently I saw I had caught his attention. He came over to my cage and asked the fancier if I, too, was part of the new broods. On hearing that I was, he said he would like to have me.

"Then, to my great delight, the fancier went and got a small traveling cage to lend his friend, until he could buy a bigger one for me. Into this I was changed and wrapping paper was put around and I couldn't see anything more after that.

"However, my parents and brothers and sisters could still hear me through the paper. They wished me good-by and good luck. Then I felt my cage lifted up off the table and the first journey of my life began.

"Of course, I wondered, inside my paper-covered carriage, where I was being taken and what my new home would be like. From the motion, a curious jerky sort of swing, I guessed I was still being carried by some one walking. But soon I was put down again, just for a moment, and the suddenly cooling air told me I was outside the house. Next I heard the stamping of a horse's feet, and then I was lifted up again, high. After that a new kind of motion began and, from the swing of it and the regular beating of hoofs, I knew I was being carried on horseback.

"I felt the wind blowing through my paper covering. Soon the scent of the ripe corn and poppies reached me and I knew that my owner was now beyond the town, out in the open country. I had never been in the country before, but my father had, when being taken to shows, and he had told me something about it.

"It was a cold ride, bumpy and uncomfortable. With the jolting motion of the saddle pommel on which my cage was held every bit of the water and seed out of my troughs got spilled all over the cage—and me, too.

"Presently the horse's pace slowed down and, hearing now the echoes of his hoofs thrown back from near at hand, I guessed that we had entered the streets of another town. I wondered if this were the place where I was to live or if my owner would ride on through it. I heard sparrows chattering, pigeons cooing, dogs barking, people talking and calling. I hoped we would stay here. It seemed a nice, cheery place, from the sounds of it.

"And sure enough, presently, to my great delight, that awful riding motion ceased. I felt my cage being handed down and taken by other hands. Somebody—a woman—was greeting my man on the horse. The air suddenly grew warm and a door slammed shut. I was inside a house—a house of many smells, most of them nice, comfortable, foody smells. My cage was set upon a table. Several voices were now chattering around me, some of them children's. Fingers began clawing at my wrapping paper, to undo it. The string was cut with a *ponk* like the twang of a guitar. And then—at last —I could see."

THE FIFTH CHAPTER

The New Home

"THE room," Pippinella continued, "in which I found myself was quite different from any I had ever seen before. But then, of course, I had only seen one other, so far—the fancier's conservatory. This place was pretty large, with lots and lots of chairs in it, a ceiling of big smoked beams and funny pictures on the walls of men in scarlet jackets galloping across country on horseback. A pair of stag's horns were hung over the door. And above the fireplace there was an enormous dead fish in a glass box.

"Gathered about my cage stood four or five people, men, women and children, all with fat, round faces and red cheeks. They were staring at me with great curiosity and—to judge from their smiles—with some admiration. I guessed them to be the family of my new owner. Presently another cage was brought and I was changed into it. It was quite roomy and decent inside and I was glad to get into it after the little crampy one which had been so messed up by the journey.

"Then there was evidently a good deal of discussion among the family about where I should be put. One pointed to one place and another to another. Finally it was decided to hang my cage in the big bay window which looked out on to a fore-court, or front yard.

"Of course, you must understand that up to this I had never seen very much of people. I was exceedingly young. At the fancier's all I ever saw, with very few exceptions, was one person at a time. But in this house it was entirely different. People in twos and threes were around, talking all the time. And, watching and listening to them the whole day, I soon began to understand many words of their language. Even that first day I guessed from signs and other things that the largest boy of the family was asking his mother if he could have the job of looking after me. Finally his mother consented—to my great sorrow later, because he was the most forgetful monkey that ever walked. Many was the time that he forget to refill my water trough and I'd go thirsty for a whole day before he found it out. He was always dreadfully sorry when he discovered his mistake, but that didn't do me any good.

"At first I was very much puzzled about this house. The family seemed to be positively enormous. All day long new batches that I had never seen before, men, women and children, kept arriving, some on horseback, some on foot, some in carriages. They would take meals in the dining room, and often sleep upstairs in the bedrooms overnight. Then they'd go away again and different ones would take their place. There was somebody arriving and somebody going away all the time.

"Then I decided these could not all belong to the family, and I supposed that my new owner was a man of many friends. Heaven only knows how long I would have gone on believing that if I had not one day had my youthful ignorance enlightened by a chaffinch. It was a warm afternoon and the bay window had been opened. I saw a chaffinch passing and repassing, with bits of horse hair in his mouth. He was busily building a nest in one of the poplars in the yard. I had not spoken to a bird since I had left my own family, so I hailed him, and he came over close to the window and chatted a while.

" 'This seems an awfully funny house,' I said. 'Who's this man who has so many friends coming to visit him all day long?'

" 'No one could mistake you for anything but a newborn cage-bird,' the chaffinch laughed. 'They aren't his friends. This isn't a private house. This is an inn, a hotel, where people pay to stop and eat. Haven't you noticed that big carriage that rattles into the yard every evening at five o'clock? Well, that's the coach from the North. And the one that comes early in the morning is the night coach from the South. Haven't you seen them changing horses? This is a regular coaching inn, one of the busiest spots in the country.'

"And very soon I decided that in my first venture away from the protection of my mother's wing I had been very lucky. Good fortune had given me a home that any cage-bird could envy. I have often looked back with pleasure upon the nice cheerful bustle of that inn. If you must be a cage-bird —if you have to be deprived of the green forests and the open freedom of the skies—then it is good to be in close touch with the world. And there one was certainly that.

"Something new was happening all the time. Men went over this road not only to the capital but to foreign lands—for it was the highway to a great port from where ships sailed to the seven seas. Travelers coming and going brought news from everywhere. And the daily coaches always delivered the newspapers from the North, South, East and West.

"All this I witnessed from my little cage. And when the summer weather came I used to be put outside every morning, high up on the wall beside the door. From there I could see down the road a long way. And the daily coaches were visible to me quite a while before they could be seen by any one else. When the weather was dry I could tell them by the cloud of dust far, far off; and then I'd sing a special song which I made up. It began *Maids, Come Out, the Coach is Here.* And though nobody understood the words of it, all the maids and the porters of the hotel soon got to know the tune. And whenever they heard me sing it they'd know the mail coach would arrive in a few minutes and they'd all get ready to receive the guests. The maids would take a last look at the dining room tables; the porters would come to the door for the valises and luggage; and the stable boys would open the yard gate and get ready the fresh horses to change for the tired ones coming in. It gave me quite a thrill on a quiet drowsy afternoon to suddenly wake that inn up to bustle and life, just by singing my song, *Maids, Come Out, the Coach is Here.* In that way, you see, I was not only in touch with the world but I was, in a manner, an active part of it. For though I lived in a cage I felt myself a responsible member of the hotel staff.

"Another thing that is very important for a cage-bird, if he is to lead a happy life, is that he shall like people. Most wild birds look upon people as just something to be afraid of and think they are all the same, like stones or beans. They're not. They're all different. There's just as much difference in people as there is in sparrows or canaries. But you can't make a wild bird believe that.

"I had not lived very long at that inn before I made a great number of very excellent friends—among the people. One I remember particularly: the old driver who drove the night coach from the North. He was what is called a very famous whip—that is he could drive a four-in-hand with great skill. Every evening when he brought his great lumbering carriage into the yard he'd call to me from his box, 'Hulloa, there, Pip!' And he'd crack his long whip with a sound like a pistol shot. Then the ostlers would all come running out to change horses, polishing up the harness, washing the mud off the traces—as busy as bees putting everything in ship shape for the next stage of the journey. I made up another song for that nice, jolly-faced old driver—his name was Jack. And every evening when he cracked his whip I'd sing it to encourage the busy stable boys at their work around the coach. It was meant to sound like the jingle of harness and the *shish-shish-shish* of a curry comb. And it ended sudden and sharp—'*Jack!—Crack!*'

"Old Jack always brought me a lump of sugar in his pocket —never, never forgot. And on his way in to get his own supper he'd poke it into the bars of my cage and take out the old one which he had left there yesterday. He was one of the best friends I ever had was Jack, the driver of the night coach from the North."

THE SIXTH CHAPTER

An Adventurous Career

AS I HAVE already said, Pippinella's life was quite a long story. And since this book is to be a history of the Doctor's adventures, I feel it would be wiser if I told the rest of the contralto canary's career in my own words, rather than in the longer form in which she narrated it herself to the Dolittle household in the caravan.

Certainly few cage-birds, indeed few people, ever experienced so many thrills in the space of a short lifetime. From the inn where she had been so happy she was bought and taken away by a nobleman who stopped there on his way to his country estate. Arriving at a very generous castle she was presented to the nobleman's wife and lived for some weeks in that lady's little boudoir at the top of a high tower.

Here she was introduced to an entirely different kind of life from what she had seen at the inn. From the grand silver cage into which she had been put, she saw trouble brewing all around her. The nobleman (a marquis) owned many square miles of land, whole towns, coal mines, factories and what not. His wife, the kind marchioness in whose boudoir Pippinella lived, was unhappy. The canary sang songs to her to cheer her up—*The Harness Jingle* which she had composed for old Jack, the driver of the night coach from the North, and the merry call, *Maids, Come Out, the Coach is Here.*

She heard rumors of riots in the factory towns and the mines. The workers were discontented. One day when both the Marquis and the Marchioness were away a mob attacked the castle, sacked it and set fire to it. The dogs and horses belonging to the estate were rescued by the workers themselves. But not so poor Pippinella. In her silver cage hung outside the tower window hundreds of feet above the ground, she was overlooked and left on the wall of the blazing building. Unable to escape, she saw the flames slowly but surely rising through the castle, floor by floor.

But just as all chance of rescue seemed most hopeless she heard the sound of drums coming up the valley. A regiment of soldiers had been sent to put down the riot and save the estate. The workers fled. The fire was put out and Pippinella, the only living thing left in the gutted building, was rescued by the soldiers.

After this dramatic escape she was made the regimental mascot and traveled with the military wherever they went. She was treated extremely well, as an individual of great importance who would bring good luck so long as she survived. From place to place she went, always riding in her cage on top of the baggage wagon, while the Fusiliers (the regiment to which she had been attached) put down riots and disturbances in various small towns round about.

During this period she composed another air—a marching song for the soldiers of whom she had grown very fond; and it began "Oh, I'm the Midget Mascot, I'm a Feathered Fusilier."

The day came when the soldiers were sent to a certain town to put down an uprising of the workers. They were commanded to fire on the crowd which was unarmed, They did not openly rebel against orders. But being really in sympathy with the workers they allowed themselves to be defeated. The baggage wagon on which the mascot Pippinella traveled was captured. And the contralto canary found herself suddenly transferred from the position of a pampered regimental pet to the possession of a laboring man who had won her in a raffle.

This man later attempted to escape from the town after it was besieged by fresh troops who had come to reenforce the defeated Fusiliers. With him he took a companion and Pippinella. On the way out through the sentry lines he was shot. And though he managed to escape and drag himself several miles into the country he finally died of his injuries and the canary passed into the possession of his companion.

Pippinella's new owner was apparently a coal miner and it was his intention to get to the next town and seek work in the coal pits. He begged a ride on a grocer's wagon and finally reached his destination.

The next chapter in Pippinella's story is a very strange and mournful one. It was apparently the custom in many mines to have canaries underground where the men worked. They were placed high up above the workers' heads on the walls of the galleries and tunnels. The idea was that the deadly gas, which is sometimes a source of great danger to the miners, would begin by gathering against the ceiling of the tunnels. Thus the behavior of these birds was supposed to give warning to the miners when they were in danger.

After some weeks of this dark and gloomy life Pippinella saw an old lady coming through the mines on a visit. She was very interested in the presence of the canary here and asked if she could buy her. The price she offered was large and the miner who owned her jumped at the chance of making so much money. The old lady took Pippinella back to her home with her and a new and brighter life began.

It was springtime and Aunt Rosie (as Pippinella came to call her new owner) decided that her canary must have a mate and rear a brood. So a gentleman canary of very smart appearance was bought at the local live-stock shop and introduced to Pippinella. She found him very stupid, she told the Doctor, but of a kind and thoughtful disposition. But the most remarkable thing about him was his voice. Pippinella considered herself (and rightly so too) a good judge of bird voices. But she assured John Dolittle that among tenor canaries she had never heard, before nor since, the equal of her first husband. His name was Twink.

Many quiet domestic weeks now followed during which Pippinella and Twink raised a lusty brood of young canaries to full growth. The youngsters were all given away, when they were old enough, to Aunt Rosie's friends. And finally Pippinella and her husband found themselves alone again.

It was at Aunt Rosie's home that Pippinella made the acquaintance of the man who had the greatest influence on her life. He was a window-cleaner. Her cage had always hung in the window. And while this man polished the panes he would whistle to the canary and she would talk back to him. Pippinella said she knew at once that he had character—that he wasn't just an ordinary person, and that he probably only cleaned windows to make his bread and butter. Aunt Rosie, though she was kindhearted, Pippinella had found a rather tiresome woman. To the canary's great delight, she was eventually presented to the window-cleaner as a free gift.

It was no very terrible sorrow to her to leave Twink, who, while he was the greatest singer on earth, had proved a very dull mate. Pippinella went off with her window-cleaner in high glee.

Her new home was a strange one. The window-cleaner lived in an old broken-down windmill. He was away most of the day cleaning windows to earn his daily bread. But he used to work far into the night writing. He seemed to be very secretive about this writing, always hiding his manuscripts in a hole under the floor when he was finished.

One day he went off as usual to his work. By this time Spring was coming again and he had set Pippinella's cage on a nail outside the window of the mill. It hung a good twenty feet from the ground. The hour came for his usual return

"Hour after hour he would sit writing"

but he did not show up. Darkness came on—and still he had not come back. Two days went by, and still no window-cleaner.

The canary's food was by this time of course long since exhausted. She felt something must have happened to her master and there was every chance, since he lived entirely alone, of her starving to death.

However on the third night, towards the morning hours, a great storm came up and her cage was finally blown off the nail in the mill-tower and sent crashing to the ground. It broke in halves. The canary was unharmed; and suddenly, as the dawn showed in the East, she found herself, for the first time in her career, *free!*

THE SEVENTH CHAPTER

Freedom

PIPPINELLA spent some time describing to the Doctor what that freedom meant. At first she was greatly rejoiced: she could go where she wished, do as she liked. But, born and bred a cage-bird, she soon found that life in the open held more dangers than comforts for one who was not experienced in it. When she attempted to make her way through the hedges like other birds she got her wings all tangled up in the blackberry brambles. She was chased by a cat, weasels and hawks. She did not know where to look for wild seed. Flights longer than a few yards tired her dreadfully. Much of her time was spent cowering in the holes of the mill-tower to escape her enemies of the air and earth.

Then came along a greenfinch. Pippinella, you will remember, was a green canary, crossbred. The greenfinch had seen her escaping from a hawk and saved her life by drawing off that deadly enemy upon himself so she could escape. In the calm beauty of a spring evening he made love to her and offered to show her how the life of the wild should be lived. Pippinella was touched. This greenfinch had saved her life. She went off with him. They were to build a nest and raise a brood.

Many leagues they traveled seeking the ideal spot for the nest which should be worthy of their romance. Often in those days she heard her lover sing that famous melody which later became well known through her, to human audiences. She called it then "The Love Song of the Greenfinch in the Spring." Meanwhile her mate coached her in all the arts of the wild life which she, as a cage-bird, was ignorant of.

" 'I got my wings all snarled up in the blackberry bushes' "

Those were idyllic days as Pippinella described them to the Doctor. But tragedy and sorrow lay ahead. One evening when she returned after seeking nesting materials (they had found the ideal place for their home in a little bay where the sea rolled in in gentle waves and the wild flowering bushes hung low down over the sandy shores) she discovered her mate in conversation with a full-bred greenfinch, a lady of his own race. Pippinella knew at once, she told the Doctor,

" 'We all three roosted on a flowering hawthorn limb' "

that the end of her romance had come. However, she tried to behave in a ladylike manner and when introduced to the greenfinch damsel she was careful to be polite and courteous. The three birds roosted for the night on the limb of a flowering hawthorn bush. But Pippinella knew she wasn't wanted —after all she was only a mongrel, a cage-bird.

Very early, before the dawn had wakened her companions, she quietly left the hawthorn bush and betook her to the shore. There on the sands she determined to fly to foreign lands, to forget—and start life anew.

By this time indeed she had already begun to weary of the wild life—of the freedom that has so many dangers for the cage-bird. She wanted to find her old friend the window-cleaner. It could not be that he was dead! And if he was alive, maybe he needed her. Her motherly instinct was aroused. Very well then, she would leave her faithless lover, with his greenfinch hussy. And scouring foreign lands beyond the sea, maybe she'd meet her old friend. With him she would return to the life of captivity in which she had been born and be useful and happy.

From this point on her life story became one long series of adventures. Crossing the ocean without a wild bird's knowledge of geography and navigation was of course in itself a foolhardy thing to attempt. Moreover she had as yet very little endurance for long-distance flying. She was no sooner beyond the sight of land than she was lost and exhausted. She took refuge on some floating gulf weed till a passing curlew gave her directions as to how to reach the nearest land.

Following these instructions she finally came to Ebony Island, a jungle-covered mountainous piece of land only a few miles square. After she had recovered from her fatigue she examined the island and found it a pleasant enough place. She was treated kindly by the native birds, many cock finches vying for her affections. But she is still too broken-hearted over her lover's faithlessness and she encourages none of them. She does not even tell them where she came from and she remains a sort of woman of mystery among the bird society of the island.

She fills in her time composing many new and beautiful songs. But after a few weeks she discovers that all her new

"'I pulled its hanging folds out with my bill'"

songs are sad—none of them jolly, like the "Maids, Come Out," or stirring, like "I'm the Midget Mascot." Wondering why this is, she decides that she is still mourning for her friend the window-cleaner. She has an uncanny and unexplainable feeling all the time when she is flying about the island that he is near her, or that he has been here before her.

At length just as she is about to leave she discovers the

"The ship came nearer and nearer"

window-cleaner's duster tied to a pole at the summit of one of the island's hills. She is quite certain it is the towel that he dried the windows with, because she knew every detail of it and it has a mended tear in exactly the right place.

Hunting around in the neighborhood of this signal station (for the cloth had been clearly set up as a flag to attract the attention of passing ships) she finds a cave where her old friend had lived and other traces of his presence.

"The Razor-strop Duet"

He must have been shipwrecked and cast away on this island, she thinks. What shall she do now? She sees a ship passing some miles away to westward. She fears her own ignorance of navigation; but following this ship she must surely be brought to land—possibly back to England where she had last seen her friend.

She sets off. But as soon as she is well out from the shores of the island a heavy rain squall drives her forward past the ship like a leaf in the wind. Her only hope of safety is to take refuge on the ship itself while there is yet time. This she does, and is immediately captured by one of the crew. She is put into a cage and placed in the ship's barber's shop. There is another canary there who tells her all the gossip of the ship.

This new life was not a bad one. She was treated well and the business of the barber's shop, where passengers came to be shaved and have their hair cut, provided a constant entertainment. During these days she taught the other canary to sing and she composed an amusing song which she called, *The Razor-strop Duet*. It imitated the *clip-clop* of a razor being sharpened and the tinkling of a shaving-brush in a lather-mug. Pippinella sang the contralto part of the duet and the other canary the soprano.

Still she was always hoping to escape, to get back to her friend the window-cleaner.

One day a great excitement is started aboard ship by the sighting of a raft. The ship's course is changed and the castaway upon the raft (a strange ragged individual with thick beard grown all over his face) is rescued and brought on board. He is in a state of extreme weakness and exhaustion and is at once put to bed by the ship's doctor. For several days neither Pippinella nor the passengers see any more of him.

But after a week he comes to the barber's shop to have his dense beard shaved off. When this is done Pippinella at once recognizes him as the window-cleaner! She whistles familiar calls to him frantically as he goes to leave the shop and he recognizes his own canary's voice.

After some argument and trial by test his ownership of

"The tramp, with a glance over his shoulder, drew nearer"

her is established and she is removed to his private cabin. There he sets to work writing his biography and Pippinella herself gets the idea for the first time of composing her own life-story in song. And it was that biography in music that she had sung to the Doctor when he first brought her home to his caravan.

On the ship's arrival at the next port the window-cleaner goes ashore to seek passage on another ship which shall take him and Pippinella back to England.

And so at last they reached their home shores and the window-cleaner at once proceeded to the windmill. He left the canary in her cage outside while he went round to the back to find a way into his old home. In his absence a tramp appeared and, hiding Pippinella's cage beneath his ragged coat, made off with her before the window-cleaner returned.

Thus, after re-finding her old master under most dramatic and extraordinary conditions, she was parted from him again at the very moment of their returning to the old place they had shared as home together for so many happy months.

After a few more changes of ownership Pippinella, the great contralto canary, was sold to the animal shop from which Matthew Mugg purchased her for the Doctor.

THE EIGHTH CHAPTER

John Dolittle's Fame

AT THIS point Gub-Gub, who had been fidgeting to say something for a long time, demanded to know what became of the window-cleaner when he left the mill the first time and how he came to be floating about on a raft in the ocean. But Jip and Dab-Dab silenced him and bade him let Pippinella tell her story her own way.

"Well, that is practically all there is of it," said the canary. "The rest is in the Animal Shop."

"Humph! And quite enough, too, I should say—for one life," muttered the Doctor, stretching his cramped hand that had written steadily at high speed for over two hours. "But there are one or two more small things I'd like to get down, Pippinella. Would you sing us that song again? I want to book the words. I already have the music written out. I mean *The Greenfinch's Love Song.*"

"Certainly," said Pippinella. "But my voice isn't what it used to be, you know."

As the canary threw back her head to sing Matthew Mugg entered the wagon, and at a sign from the Doctor for silence took a place at the table and prepared to listen.

When the little contralto prima donna warbled in a whispering tremolo the opening notes of that short but thrilling melody, the whole company was instantly spell-bound. At no point did the song, as Pippinella rendered it, reach full voice. Throughout it was subdued, caressing, almost like some one humming a lullaby a long way off beneath his breath. Then it seemed as though the singer were moving still further away, hunting, searching, seeking through enchanted forests—now hopeful, now sad again; now distant, now near. It was all the mystery and beauty of the world packed into a little crooning tune. Quieter yet it grew, softer and further away, and as it faded it rose in pitch and finally died out on a top note in a muffled but bell-like trill.

Then for a few moments there was complete silence in the wagon.

"Oh, my," gulped Gub-Gub at last. "Isn't it wonderful! It reminded me of dew-spangled cauliflowers glimmering palely in the moonlight."

Matthew Mugg, the cross-eyed Cats'-Meat-Man, who always said he did not care for music, turned to John Dolittle.

"Doctor," he said, "nobody ain't heard nothing like that never before. That bird's a marvel. And by Jiminy, why don't we make *this* the turn for London?"

"That was what I planned to do, Matthew," said the Doctor. " 'The Canary Opera.' I have a feeling it will be the best show we ever put on—the most artistic animal performance ever seen. It is all here. The story of the opera will be Pippinella's own life. You couldn't have a better libretto. The prima donna will be the lady herself. We will take London by storm."

"Oh, not a doubt of it," said Matthew, blowing out his chest. "We'll bowl 'em right over. We'll put all the other opera-mongers out of business."

"Of course," the Doctor went on, "the details, such as choruses, orchestra, scenery and costumes, we can work out later. But the main things are here: Pippinella's voice and Pippinella's story. In them we have the makings of a great performance."

"But look here," squealed Gub-Gub. "You're always blaming me for interrupting. Now you're all doing it. Pippinella hasn't finished yet. I want to know how she came to be in the animal shop where the Doctor bought her."

"It reminded me of dew-spangled cauliflowers"

"You're quite right, Gub-Gub," said John Dolittle. "Pardon us, Pippinella. I was carried away. Matthew's remark about the London performance set me going. Please proceed with your biography."

And as he sharpened his pencil and turned over a new page in his notebook, Pippinella got back on to the tobacco-box and prepared to continue the story of her life.

"It often seems sort of queer to me when I look back on it," she went on, "that I had never heard of you before, John Dolittle. Of course, if I had been a regular wild bird I could not have helped it. But you must remember that, while I had led a very eventful life, it had been spent more among people than among animals. Even when I had my liberty I did not, as I told you, mingle much with other wild birds. Still, it is curious, none the less, that I had not at least heard your name.

"Well, one day in the animal shop while I sat dejectedly on the one dirty perch which our cage boasted of, thinking of the window-cleaner—wondering, as usual, how he was getting on—I heard the other birds in our cage conversing.

" 'Just look,' said one mangey hen to another, 'at that miserable thrush across the shop there. The little box they've put him in is so small he can scarcely turn around without crumpling his tail against the wall.'

" 'Yes,' said another. 'And the poor blackbird next to him is worse off still. His place hasn't been decently cleaned out since he came here. He's ill, too.'

" 'I wish,' said the first, 'that the Doctor would come. I'm so sick of this miserable place.'

" 'And if he ever did,' said the other, 'why do you suppose he'd pick you out of all the beasts, birds and fishes gathered here? One can't expect him to buy the whole stock!'

"Then another scrubby runt who was eating at the seed-trough joined in the conversation with her mouth full.

" 'I'll tell you what we ought to do,' she piped. 'When John Dolittle comes, since he can't buy us all, we ought to ask him to set up a canary shop that's managed properly.'

" 'And what kind of a shop would that be?' asked the others.

" 'Well, for one thing,' said the little hen, 'it should only

" 'Yes,' said another, 'and the poor blackbird is worse off still' "

have the kind of cages that we approve of and proper, decent attendance for every bird in it. But, most important of all, the birds should be allowed to select their own buyer. If a customer came in whose face you didn't like you wouldn't have to be sold to him.'

" 'Oh, what's the usc of talking,' said the first one. 'The Doctor won't come. I've been in this place for over a year and he's never shown up. They say he's afraid to come to animal shops—hardly ever goes near them.'

" 'Why?' asked the little hen.

" 'He can't bear to see the animals and birds badly cared for. They all yell at him as soon as he appears, imploring him to buy them. Of course, he can't. Hasn't got enough money—never has any money, apparently. And, understanding the language of animals as he does, he is merely made miserable by the visits without doing any good.'

" 'Well,' said the little hen, 'you can't tell. He *might* come some day. And when he does I'm going to ask him to buy me.'

"So far I had taken no part in the conversations. I had felt too unhappy for gossip, to tell you the truth. Hadn't opened my mouth, hardly, since I had been put in. But this talk interested me.

" 'Who is this doctor?' I asked.

"They all looked at me in surprise.

" '*The* Doctor,' said one. 'Doctor Dolittle, of course.'

" 'And who is he?' I asked.

" 'Great heavens!' said the little hen. 'Can it be that you have never heard of him? Why, he is the only real animal doctor living. Talks all languages, from canary to elephant. I didn't know there was any creature left in the world who hadn't heard of him.' "

THE NINTH CHAPTER

The Show for London

"AND after that I very soon discovered from bits of odd conversation that drifted to me from all parts of the shop that every creature there, John Dolittle, had one hope always in his mind, and that was that some day you would walk in and buy him."

"Alas!" murmured the Doctor. "I wish I could. But what your neighbor told you was true: I just dread to go near an animal shop."

"Well," continued Pippinella, "my long story is nearly ended. After that I, too, joined the band of hope. Day after day and week after week I watched the door, like the rest, to see you come in. From the other birds I made inquiries to find out exactly what you looked like, so that I could recognize you at once, although I had never seen you before, if by chance you should some day come.

"Later I was given a separate cage and put in the window. And one day—I shall never forget it—I saw your high hat go past the shop. I made all kinds of signals, but you were not looking. You hurried on, afraid to be seen. But you had already been recognized by half the creatures in the place. I shared the general disappointment when you passed on without entering. Then the man with you came back. At once the news was shouted from cage to cage and pen to pen that he was your assistant and was making a purchase for you. But of course he didn't know bird talk and I couldn't make him understand the message I wanted carried to you. But at last you yourself reappeared and looked in the window. I shouted as loud as I could, but was unable to make you hear through the glass. You started to go away again.

" 'I saw your high hat go past the shop' "

I fluttered and made signals. I saw that you noticed me. But still you went away. A terrible feeling of despair came over me. And then your assistant reentered the shop. You can't imagine my delight when I saw that he was pointing to my cage and telling the proprietor that he wanted to buy *me*. The rest of the story you know. I was sorry for all those poor creatures I left behind—the ones whom you didn't buy. But, oh, goodness! I was glad to leave that smelly place and get back once more among people who treated me like a friend, instead of just as something to be sold."

There was a little silence after Pippinella stopped speaking—the only thing audible was the scribbling of the Doctor's pencil as he wrote the last words in his notebook of the Contralto Canary's Autobiography.

"Well," said he at last. "It's a wonderful story, Pippinella. You have been through a great deal. Some day, who knows? when we have made more money with the circus, I may yet set up the Canaries' Own Bird Shop along the lines suggested by your cage mates whom you had to leave behind. There is no reason why cage-birds should not be allowed to pick their own buyers. I think your friend's idea is a very good one. We must see what can be done. And now I have a favor to ask of you. . . . Where did Matthew go, Jip? I didn't see him leave the wagon."

"I think he's over at the menagerie, Doctor," said Jip.

"All right," said the Doctor, turning to Pippinella again. "If you will wait here a minute till I go and fetch my assistant, Mr. Mugg, I would like to put a request before you for your consideration."

"Oh, dear!" sighed Gub-Gub, as the Doctor left the wagon. "I'm so sorry the story's over."

"But it isn't over," said the white mouse. "Pippinella's story isn't finished while Pippinella is still alive. We've only heard the part that belongs to the past. There still remains that part which belongs to the future. After all, it's much better to feel that you are living a good story than that you are just telling one."

"Yes," said Gub-Gub. "I suppose so. Still—you can make it so much more exciting when you're only telling it. Myself, for instance, when I have day dreams about food—twenty-

" 'You have a romantic soul,' growled Jip"

course meals, you know—they are so much better than the lunches and dinners I ever get in real life. Real life, I find, is not nearly so thrilling—seldom runs to more than boiled beef and cabbage and rice pudding."

"You have a romantic soul," growled Jip. "Food—always food!"

"Well," said Gub-Gub, "there's lots of romance in food if you only knew it. Did you ever hear of Vermicelli Minestrone?"

"No," said the white mouse. "What is it—a soap?"

"Certainly not," Gub-Gub retorted. "Vermicelli Minestrone was a poet—a famous food poet. He married Tabby Ochre. It was a runaway match. But she stuck to him through thick and thin. People said she was a colorless individual and would stick to anything. But he loved her dearly and they were very happy. They had two children—Pilaf and Macaroni. He was a great man, was Minestrone. His library consisted of nothing but cookbooks—cookbooks of every age and in every language. But he wrote some beautiful verses. His Spaghetti Sonnets, his Hominy Homilies, his Farina Fantasies—well, you should read them. You would never say again there was no romance in food."

"It's a sort of cereal story," groaned Jip; "mushy—Ah, here comes the Doctor back, with Matthew."

A moment later Manager Dolittle, accompanied by his assistant, Mr. Mugg, reentered the wagon. With an air of business, they immediately closed the door and sat down at the table.

"Now, Pippinella," said the Doctor, "I want to talk something over with you which you have already heard Matthew and myself refer to. It is the Canary Opera. The story of your life, one of the most interesting accounts I have ever listened to, would make an excellent libretto for a musical production of an entirely new kind. We have, as you know, been invited to come to London and perform there. We want to give them something new and something good. The Canary Opera strikes us as just the thing. You will, of course, take the leading part—the heroine, the prima donna. We would have to have a cast of singing birds to support you—especially good voices for the principal rôles and many others to form the choruses. Will you help us? Would you be willing to go into this?"

Pippinella put her head on one side and thought a moment.

"Why, of course," she said at last. "I'd be delighted."

"Splendid!" cried the Doctor, "splendid! Then in that case, Matthew, I think we will cancel the rest of our engagements with the small towns and get under way for London as soon as ever it can be managed."

THE TENTH CHAPTER

On the Road

GREAT was the excitement among the animals of the Dolittle household when it became known that the circus was setting out for London earlier than had been planned. Even the thrilling weeks in Manchester (of which Gub-Gub, for one, never ceased to recall the details) took a second place when compared with this invitation to visit the capital.

"It's the biggest city in the world," the white mouse kept saying. "The Dolittle Circus is becoming a pretty important concern, you may be sure, when they send for us especially to come all this way. Maybe the Queen and the people of the Court will come and see the show."

Jip, who had been there before, could not help boasting of his knowledge. And the others would sit around him by the hour while he related the wonders of the great town and answered the hundreds of questions which were put to him.

Matthew Mugg left that same night. And, with a letter for the London managers in his pocket, he set out to make arrangements ahead for the coming of the Dolittle Circus. The Doctor felt it would be as well to put the Puddleby Pantomime in rehearsal again and to play that in London as well as the Canary Opera. And then, if at the last minute it should prove impossible to put on the opera, the pantomime could be made the principal turn.

So Gub-Gub, Dab-Dab, Too-Too and the three dogs, Jip,

Toby and Swizzle, filled up the few days of waiting by going over their parts and dances and making sure they remembered the pantomime without a hitch from beginning to end.

Much to Gub-Gub's dismay, he found that his figure had greatly altered since he was last on the stage. He had grown much fatter. He could hardly get into his clothes at all, and at the first rehearsal buttons kept flying off him in all directions and seams gave way with loud cracks. What was worse, he found that walking on the hind legs for him was now almost impossible. With all this weight in front, he was top-heavy.

"What shall I do, Doctor?" he asked, almost in tears.

"Well," said John Dolittle, "if you want to play in the pantomime there's only one thing to be done: you'll have to diet."

"To *dye* it!—To dye what?"

"No, no. To *diet,*" the Doctor repeated. "That is, to eat only special things, things that don't make you fat."

Gub-Gub's face fell.

"What things?" he asked.

"You will have to give up vegetables," said the Doctor. "Stick to rice and that kind of thing."

"Give up all vegetables?" asked Gub-Gub: "Parsnips? Potatoes? Turnips? Beets?"

At each one the Doctor shook his head.

"What is life without vegetables?" asked Gub-Gub tragically.

"To beet or not to beet," Swizzle whispered into Jip's ear. "To diet—perchance to dream—"

Swizzle, the old circus dog, remembered a good deal of Shakespeare which his master, Hop the Clown, used frequently to misquote in the ring.

"Oh, well," said Gub-Gub. "I suppose it has to be. The

things we actors have to give up for our art! Maybe I can make up for it after the show is over."

Three days later Matthew Mugg rejoined the circus and told the Doctor that matters had been satisfactorily arranged

" 'What shall I do, Doctor?' "

in London. A camping ground had been booked and the managers, while impatient to have the Doctor's show as soon as possible, were quite willing to wait till he had got every detail the way he wanted it.

Then began the big business of packing up the circus. That, of course, was no extraordinary thing for them, whose life for so long had been one of continual shifting from town to town. But this occasion was a very special one, and on this trip they planned to do more miles per day than usual, so as to get to London as soon as possible.

The first real stop was to be Wendlemere. And the Doctor intended to get his own wagon there in one day, if it could be done with a change of horses. But, of course, sixty miles without a night's rest was too much for the others. Hercules, the strong man, was to bring on the remainder of the train the following day. Matthew was to travel in the Doctor's wagon.

So very early in the morning, before the rest of the circus was barely awake, Manager Dolittle's caravan, drawn by the fastest horses that the stables could provide, set out along the road to Wendlemere. About half way they changed their horses at a village inn and took on fresh ones, which had been sent ahead of them the day before. The old ones were left behind, to get a good rest and be picked up by Hercules on the morrow. All these details had been arranged by Matthew Mugg, who at this part of the circus business, was as good as three men rolled into one.

With the fresh horses they reached Wendlemere easily a little after dark and pitched camp for the night just outside of it, on a strip of turf bordering the road.

"This town," said the Doctor, as Dab-Dab began to lay the table for supper and Jip to scurry around the hedges after sticks for the fire, "is famous for its buns. The Wendlemere buns are almost as well known as the Banbury cakes or the Melton Mowbray pies."

"Humph!" said Gub-Gub. "Did you say buns? How would it be if you should send Matthew in to get us a few, Doctor? It's only just dark and the shops will still be open."

"Buns," said the Doctor, "are very bad for the figure. If you want to play the part of Pantaloon, Gub-Gub, you will have to leave pastry alone for the present. It's dreadfully fattening."

"Oh, dear!" sighed the pig. "Vegetables and pastry and everything off the list! All I get is boiled rice and thin soup. Dear, dear! How I shall eat when the show is over! It's funny how some towns get well known for certain kinds of food, isn't it? You know, I heard of a place once which became famous for second-hand joints. Yes, they had a whole lot of hotels there which always wanted new, uncut joints on the sideboard. So as soon as a sirloin of beef or a saddle of mutton had been used for one meal it was sent back to the second-hand joint merchant, to be sold over again. It's curious."

"Very skewerious, I should say," muttered Too-Too. "You're frightfully well informed—on food."

"Yes, indeed," said Gub-Gub proudly. "I intend to write a book on it some day—'The History of Eating.' "

THE ELEVENTH CHAPTER

The Caravan Reaches London

IN TALKING over with Pippinella where good bird voices for the other principal parts in the opera could be obtained, the Doctor had agreed with her that if they could only get her husband, Twink, the leading tenor rôle would be very well filled. There was not much hope that he could be found. But the Doctor had before leaving made a special visit to Aunt Rosie to see if he could borrow Twink for the opera. He was told that the great tenor had been given away. Following the scent still further, he found the bird had next been sold to an animal shop in Wendlemere. It was for this reason that the caravan was proceeding to London by way of this route, which was not as short as the one usually followed by the stage coaches.

The next day the Doctor, taking Matthew and Pippinella with him, set out for the animal shop in the town. John Dolittle himself did not venture inside, but sent Matthew, with Pippinella in her cage beneath his arm, to find out what they could about Twink.

The proprietor seemed a decent enough little man, anxious to give any information he could. He was unable, however, to remember selling the bird which the Cats'-Meat-Man described. He said he sold in the course of six months a great number of singing birds and it was quite impossible for him to remember this one among so many. Then Matthew and Pippinella went around all the cages, to see if by chance

Twink still remained in the shop. It had been arranged that Pippinella should give a loud whistle if she recognized her husband, so that Matthew should know which one to buy. But, as she had suspected, he had gone. Pippinella questioned several of the other birds, and they said they remembered him quite well. He had been sold, they thought, to a man who seemed to be a dealer from some other town—because he had bought up quite a number of good singers, Twink among them. But where he came from they did not know.

There seemed nothing further to be done. So they left the shop and rejoined the Doctor down the street, and Pippinella reported the results of their visit.

"Well," said John Dolittle, "it's too bad he's gone. But I do not give up hope. I shall have to send Matthew to several shops to get some of the birds we will require for the opera, and it is just possible that he may run across your husband in the course of making his purchases. I will have you go with him, so that you can help him."

Then they proceeded back to the wagon, where they found breakfast all cooked and laid out for them by the thoughtful Dab-Dab. After they had eaten they waited where they were for the rest of the circus to join them. And a little after noon the first caravan, in charge of Hercules the strong man, came in sight down the road.

The remainder of the day was spent in resting up the horses for the next stage of the journey and in rehearsing the Puddleby Pantomime. Toward evening Gub-Gub suggested that the Doctor take him and the dogs for a walk. He was anxious to get his weight down as soon as possible, and he preferred doing it by exercise, he said, rather than by too much of this dieting, which was not at all to his liking.

So John Dolittle took them all for a brisk walk along the country roads. And Jip, Swizzle and Toby put up a hare out of a hedge and gave him a fine cross-country run—pretending they didn't hear the Doctor calling to them to leave the poor thing in peace. Gub-Gub, bent on reducing his figure, joined in the chase. But he had a very hard time keeping up

"Poor Gub-Gub was chased all the way home"

with the dogs and darting through the small holes in the hedges. And he hadn't gone across more than a couple of fields before an enormous sheepdog belonging to a farmer joined the hunting party and took the pig for the quarry instead of the hare. Poor Gub-Gub, who had set out hunting, was chased in a breathless state of exhaustion all the way home to the wagon, where he told the Doctor that, on the whole, he thought dieting had advantages over exercise as a means of reducing the figure.

Theodosia, Matthew Mugg's quiet but very useful wife, kept what she called the Diary of the Dolittle Circus. Her ordinary duties were light, those of wardrobe woman, hostess at the regular afternoon tea and general helper. In odd moments, when she wasn't sewing on buttons for Gub-Gub or wrapping up the packets of peppermints for the children, she took great pleasure in setting down the daily doings of the show. Her mastery of the arts of reading and writing were greatly and constantly admired by her husband, who was not, as you know, very well educated. And later, as a matter of fact, Theodosia's diary proved itself useful to John Dolittle himself on more than one occasion, when he wanted to know on what date his circus had visited such and such a town or some other detail in his life as a showman.

Well, in due time the caravan reached the great town of London (amid terrific excitement among the Dolittle animal household). It was met by Matthew and was at once set up on the camping-ground he had rented for it, on Greenheath, just outside the city. And that day Theodosia entered in her journal as a red-letter date, marking the beginning of preparations for the Canary Opera, the work which John Dolittle considered the greatest achievement of his career as a showman.

Without delay the Doctor turned his attention to the problem of getting chorus birds for his show. And the first thing he did was to consult Cheapside. He felt that that wise bird, so thoroughly familiar with all the resources of his native city, could give him the best advice on the birds available within the limits of London.

"What I need, Cheapside," said he, "is, firstly, some good

"Sewing buttons on Gub-Gub"

canary voices for the principal parts, and secondly, a great number of birds for the choruses."

"What sort of birds?" asked the sparrow.

"Well, that's just the point," said the Doctor. "There are

several choruses. I have not yet decided on what they shall be. I thought I would find out from you what would be possible here in London."

"Humph!" said Cheapside thoughtfully. "I reckon the first thing to do would be to go around the Zoo and take a look at the collections. You won't be able to buy any there, of course. But, then, after you've picked out the kinds you think suitable, I'll see what can be done in the way of supply. O' course, for good singers we'd better try the bird shops."

"Well, that Matthew will have to do," said the Doctor. "But I think what you say sounds sensible. And if you are at liberty this afternoon I'll take a run up to the Zoo with you and see if we can decide upon the best kinds of chorus birds for our show."

"Right you are, Doc," said Cheapside. "I ain't got nothing special to do to-day. The missus asked me to get 'er some greens for this evening, but I can pick them up in Regents Park. I'll be glad of the change. Ain't been to the Zoo in weeks. You know, I used to live there once."

"What, inside the Zoo?" asked the Doctor.

"Oh, not in the cages," said Cheapside. "But I was one of the Regents Park gang when I was a nipper. It's a good place for sparrows—quiet, except Sundays and Bank Holidays. I used to 'ang around the Zoo enclosure a good deal. I know the place inside out, keepers and naturalists and bobbies and all. I'll be glad to take you over it."

"Good!" said the Doctor. "Let us have lunch and get on our way."

THE TWELFTH CHAPTER

Voice Trials

ON THEIR way to Regents Park John Dolittle and Cheapside were joined by Becky, the cockney sparrow's wife. They all began by taking tea, when they had arrived within the Zoological Gardens, at one of the little tea-houses that catered to visitors. Here the two birds, sitting on the table beside the Doctor's cup, stuffed themselves with cake-crumbs and entertained him with many amusing stories of the days when they had lived here in the Zoological Gardens.

After their tea was finished they started out on their trip around the Zoo. Although John Dolittle did not intend to visit more than the bird collections, he had in the course of his inspection to pass many cages and enclosures for animals. These so interested him that Cheapside had the hardest work to get him way from them. And Becky said more than once that it would surely be dark before the Doctor had seen all the birds.

"What is that little creature?" the Doctor asked as his guides tried to hurry him by a pen containing some slim furred animal.

"Oh, that's the Genet," said Cheapside. "Nifty-looking little cove, ain't 'e? See them smart stripes down 'is sides? 'E always reminds me of some trim old maid what wears 'er 'air just so. Can't abide 'avin' 'is cage untidy. 'E 'as tidiness on the brain. 'E's kept awful busy poking the peanuts back through the wire what the children push into 'im. 'E doesn't eat 'em. But that don't make no difference to the folks what comes to Zoos. They think that everything in menageries eats peanuts. They're a bright lot, the public. They'd feed peanuts to a marble clock, they would. Yes, poor Mr. Genet spends 'alf 'is day tidying up 'is cage after the public 'as left, and the other 'alf brushing 'is 'air and cleaning 'is nails."

"Humph!" said the Doctor. "Not a very exciting life. Still, I suppose it's something to occupy the time. It must hang heavy in a Zoo."

"A feller like the Genet would find something to do in the Sahara Desert," said Cheapside—"probably tidying up the sand there. Now, the next cage, Doctor, is the Laughing Kingfishers. They're light-'earted birds—guaranteed to liven up a funeral. 'Ow would they do for choruses, do you think? Fine, strong voices, they 'ave."

"Humph! They're not very musical, are they?" said the Doctor as a dozen of the queer birds suddenly burst into hearty roars of merriment to show they appreciated the great man's visit.

"Yes, but they'd be fine for the comics," said Cheapside. "Was it a comic opera you was contemplatin'?"

"Er—not exactly," said the Doctor—"though, to be sure, there will be some comedy parts in it. But there the laughter ought to come from the audience, not from the stage. Let us go across to the big aviary over there."

This, when they reached it, proved to be a very beautiful place. It was a cage a good thirty feet high, fifty long and forty wide. Within, quite large trees were growing, and it had a pond and rocks and plants and everything. A great number of different birds were flying around or dabbling in the water or settling in the boughs. There were flamingoes, herons, gulls, ducks—birds of all sizes, shapes and colors. It was a very pretty picture.

Picking out the ones whose appearance for stage purposes struck him as suitable, the Doctor called them over to him in turn and conversed with them through the wire. He got all of them to sing a few notes for him, so as to try out the pitch of their voices. And Mr. and Mrs. Cheapside had hard work concealing their laughter when huge birds that never sang a note in their lives opened their big bills and let out strange, deep grunts and gurgles.

"Lor' bless us, Becky!" tittered Cheapside, holding his wing over his face. "Seems to me as if the Doc would do better with Gub-Gub as a barytone. If it was me, I'd give some of these blokes a job as fog 'orn."

Before they left the aviaries and started back for Greenheath the Doctor had practically decided on two choruses at all events: one of pelicans (bassos) and one of flamingoes (altos). Cheapside said he knew of a rich man living on the other side of London who had a magnificent private park given over entirely to birds and waterfowl. The sparrow added that he would go there first thing to-morrow morning to see if there were any pelicans or flamingoes in the collection.

The next day he came back to the caravan with the news that fifteen fine pelicans lived in the private park belonging to the rich man he had spoken of. And as soon as the Doctor had an opportunity to get away he went and called upon their owner and asked permission to borrow them. It happened that this man, although he was very wealthy, was a naturalist himself. Rare birds and orchids were his two hobbies. Finding in John Dolittle a scientist after his own heart, he showed him over his whole estate. And the Doctor spent a most enjoyable afternoon inspecting acres of greenhouses filled with gorgeous orchids and touring through the immense private park where a great number of birds lived happily in a semi-wild state. He was able to give the rich naturalist a great deal of information and advice on the proper preparation of thickets and ponds and nesting retreats for his different species of birds.

The Doctor was still full of all he had seen when he returned to his caravan in the evening.

"My goodness, Dab-Dab," said he as he sat down to supper, "that's my idea of a nice life—being able to spend all the money you want on your hobbies. Most rich men fritter their money away on theaters and cards and all manner of stupid stuff. Wealthy scientists are rare. That man deserves a great deal of credit for spending his time and his fortune on natural history. I've never wanted to be rich, but, by George, I came near to wishing it this afternoon when I saw the perfectly lovely place that man has up there."

"If you had been in his shoes you would have spent the fortune twenty-five years ago," said Dab-Dab sourly. "I sometimes think you should have been married—and had a strict, economical wife."

"But what would have been the good of that?" said John Dolittle. "Then I wouldn't have been able to do anything I want. Well, now, cheer up, Dab-Dab. Maybe we'll make so much money with the Canary Opera that not even I will be able to spend it all."

"I used to hope for that—once," said Dab-Dab, gazing sadly out of the window. "But you've made so many fortunes and lost them again. Puddleby and the dear old house and garden have grown to be just a dream—just a dream.—Heigh ho! What's the use? I suppose we'll be traveling circus folk now for the rest of our days."

There were real tears in the housekeeper's eyes. For months and months now she had been hoping that the Doctor would turn his face homeward and get back to the little house with the big garden. But always something new turned up in this showman business which kept putting off his return.

"Oh, come, come!" said the Doctor, genuinely touched by the tone of bitter disappointment in his old friend's voice.

"There were real tears in the housekeeper's eyes"

"It's never too late to change. The sight of that man's place this afternoon set me thinking of my own garden—it was similar in many ways—a place where a man has tried to work out the chief interests of his life. It made me long for a sight of that old wall that overhangs the Oxenthope Road, Dab-Dab, really it did. And listen: I'm going to make a real effort this time. I think the opera's going to be a huge success. That naturalist will lend me the pelicans and flamingoes I want. We'll make a fortune, Dab-Dab, and then—*then* we'll go back to Puddleby!"

THE THIRTEENTH CHAPTER

The History of Bird Music

SHORTLY after that the pelicans and the flamingoes from the rich man's park arrived at the Dolittle Circus. A special runway with a pond and thickets of bushes had been prepared for them, and, although it was of course much more cramped for room than the spacious park they had left, they declared it was very comfortable and said they were quite content to put up with some inconveniences to oblige John Dolittle. There were a dozen birds of each kind. The public that visited the circus inclosure daily took them for part of the show. And, indeed, the strange appearance of the pelicans and the handsome figures and coloring of the flamingoes, stalking about within the pen, presented quite a fine sight.

For the present the Doctor's intentions were only to get them used to seeing crowds of human faces, so that they should overcome their natural shyness, and to drill them in what they were expected to do on the stage. The singing rehearsals would have to wait till all the other birds had been collected and the musical score of the opera had been written out in every detail by Pippinella and Manager Dolittle.

The Doctor had a stage rigged up inside the birds' fence and every day he put them through their paces, showing them exactly how they were to walk on the stage, bow and take their places.

In everything belonging to the opera the inquisitive Cheapside took the greatest interest, and during these days he followed the Doctor about wherever he went. This John Dolittle was very glad of, because the shrewd little sparrow

"The Pelican chorus"

often came in extremely handy. So quickly did he catch on to the idea of drilling a chorus that quite early in the proceedings the Doctor turned the entire duties of chorus-master over to him. And then a new side-show was added to the other unusual features of the Dolittle Circus. At four o'clock, the rehearsal hour, every afternoon the fence about the pelicans' enclosure was black with the crowds watching the small cockney sparrow drilling his enormous performers. The Doctor said it was a good thing that the public didn't understand bird language, otherwise Cheapside's dreadful swearing when any of the clumsy chorus birds made a false step or got out of place would surely have caused a great scandal.

Sitting upon the top of a bush, with his chest puffed out, the tiny chorus-master looked like some angry sergeant drilling a squad of awkward recruits.

"Now, then!" he would yell. "Troop on, all in step, chins up and smiling at the haudience.—No, no. That ain't a bit like it. Any one would think this was Monday morning in the police court, instead of Garlic Night at the opera. This is *Opera,* you understand, Grand Opera. Look 'appy, not guilty! Clear off again—all of yer! Lor' bless me! Look at Mrs. Bandylegs there, thinking over the gas bill! Cheer up!—Smile! Come in, trippin'—light-'earted, not 'eavy-footed. Now, then, once more: When the music strikes up that's your cue to come on. Now—*lah, tah, tiddledy tah!"*

The next task for the Doctor was to gather together the rest of his cast for the opera. And as soon as he had turned over the training of the pelicans and flamingoes to Cheapside he took the first opportunity that presented itself in his busy life for a day in the country. On this expedition all of his animals accompanied him—even Gub-Gub, who so frequently of late had complained that he was left out of all parties. And, with the whole of his strange household following at his heels, John Dolittle betook him to the pleasant fields where, over an excellent picnic luncheon provided by Dab-Dab, the wild birds of the hedges and the woods gathered about him and sang him their songs.

On this expedition into the country Pippinella of course accompanied the Doctor to help in the judging of the voices. John Dolittle said afterwards that he could not recall any outing that he had enjoyed more. It was one of those days late in the year when summer seems to try to come back again. The sun shone without being unpleasantly hot, and all the birds of the countryside that had not yet departed for the winter migration flocked around, anxious to be accepted for the Doctor's great experiment in bird music. Indeed, John Dolittle, experienced as he was, learned a great deal that day about the songs of birds and the history of their various melodies.

After trying out different birds for the treble chorus, both the Doctor and Pippinella were best of all impressed with the common thrush.

"Tell me," said the Doctor, as a fine cock finished off a wonderful melody, "that is what you call your Evening Song, is it not? Now, has that song always been the same? I mean, have all thrushes sung it just that way always?"

"Oh, no," said the thrush. "But it has been sung pretty much that way for nearly seven hundred years. In medieval times it was quite different. Musically speaking, fashions were very much stricter then. For instance, in the song I've just sung you the middle crescendo passage was different. We ended on the major then, not the minor. Like this, *Toodledu—oo—too—tu!* instead of *Toodledu—du—tee—too!* About the thirteenth century a good many fine singers rebelled against several of the old musical rules, including the one forbidding consecutive fourths in the major scale and sevenths in the minor. That was around the time of the Magna Charta. Everybody was rebelling against something then. They didn't allow accidentals in melodies before that, either. Now we just throw them in regardless. But, in the main, the Evening Song of the Thrush is usually sung pretty much the same way now. It is in the little phrases, something between a call and a song, that you can tell whether a bird

" 'That is what you call your Evening Song, is it not?' "

is a good composer or not. Because those are what he just makes up out of his head on the spur of the moment."

"You mean when he gets an idea which he thinks will sound pretty?" asked the Doctor.

"Yes," said the thrush. "For instance, when he sees a particularly fine sunrise and tries to describe it in a snatch of music, or when a thought comes into his head about the mate he spent last spring with."

"Good!" said the Doctor. "Now, I see that you yourself are quite a musician and I would greatly appreciate it if you would do me a favor. People's notions on music and birds' ideas are somewhat different. My plan in the Canary Opera is to show people what birds can do—musically. But in order to make the score understandable for people I shall have to do certain things. I want you to compose the thrushes' chorus for me—the treble chorus which comes in the middle of the second act—when it is raining. I notice that thrushes always sing their best just when the rain is stopping. I will give you the words of the song later. I want you to get plenty of rain into your voices—it must describe the joys of rain from the thrushes' point of view. I would like you to get a choir of about twenty thrushes together and rehearse them yourself. And please be sure that the birds keep together—that they all sing the same parts of the song at the same time. I know this is not important in bird music, but it is very important in people's music. I will send you the words to-night and I'll come back here in a week's time to see how you are getting on. Can you do that for me?"

"Why, certainly," said the thrush. "I'll set about it right away."

THE FOURTEENTH CHAPTER

The Finding of Twink

A FEW days later the Doctor again went out into the country to see how the thrushes' choir was getting on. He heard the birds sing their Rain Chorus and was very well pleased with it.

"The next thing," he said to Matthew when he got back to the circus, "is to collect some good canaries for the principal parts. We shall need about five or six."

"Why, I thought there was only to be three altogether," said Matthew—"Pippinella, the tenor and the baritone."

"No, you've forgotten her mother and father," said John Dolittle. "Besides, we shall need understudies as well."

"What's understudies?" asked Matthew.

"Understudies are extra actors," said the Doctor, "who learn the parts as well. So that in case any of the principals fall sick we have another to put in his place. But they won't all be canaries. We shall need four canaries and three greenfinches. But all must be the very best singers possible—no matter what they cost. I want you to do this purchasing for me, and I will send Pippinella with you in her traveling cage so that she can help you judge and try out the birds' voices properly. I will arrange with her to give you some kind of signal when you find any that she considers especially good."

"All right," said Matthew. "When shall I go?"

"I think you had better start off first thing to-morrow,"

said the Doctor. "Time is getting short. I have promised the theater owners to have everything in readiness to open by the second week of next month."

So on the morrow Matthew went off, bright and early, canary-hunting through the bird shops of London with Pippinella. And on his return in the evening the Doctor was delighted to see him carrying another cage under his arm beside the prima donna's little traveling cage.

"He was the best we could find, Doctor," said Pippinella as John Dolittle unwrapped the paper and discovered a neat little yellow and black cock canary within. "He has a fine voice and I think you will like him. But it's slow work. You'd be surprised how hard it is to find really good singers. And as for singing greenfinches, they seem to be scarcer than diamonds. We went to over a dozen shops. But we haven't covered half of London yet. We hope to do better to-morrow."

The Doctor was highly pleased with the new member of his opera company. And that night he began practising the duets of the first act which the two canaries would sing together.

On the morrow Matthew and Pippinella went forth again and this time when they returned the Doctor could hear the prima donna calling to him even before the Cats'-Meat-Man reached the manager's wagon.

"Doctor," she cried, "Doctor, listen!"

"What is it?" asked John Dolittle, jumping up from the table and coming to the door.

"What do you think?" said Pippinella. "You'd never guess. *We've found Twink,* my husband, after all! And we've got him with us."

"Well, well!" cried the Doctor. "This is a great piece of luck. Fancy your finding Twink after all this time! Uncover the cage, Matthew. I am most anxious to see him."

"Yes, and it was only by the merest chance we did it, Doctor," chattered Pippinella excitedly. "We came to the dirtiest little shop away down in the East End of London. Matthew wasn't going to go in at all at first, it looked such

"He began practising the duets of the first act"

a poor, mean place. But I thought to myself that if there *were* any good singers there, it would be a mercy to rescue them from such a disgusting home. So in spite of Matthew's not knowing any bird talk I did my best to make him see what I wanted. And finally, after I had fluttered all over the cage and squawked whenever he turned away from the shop, he saw what I meant and went in. With every shop that I had been taken into so far I began by asking in a loud voice, 'Are you here, Twink?' But in this one I was so appalled by the dirt and the misery of most of the creatures confined there that I had not uttered a word. But I was hardly within the door when I heard a canary crying from the back of the shop. 'Pippinella! Pippinella!—It is I, Twink!—Come over here and talk to me.'

"But it was no easy matter to come over there and talk to him. I don't know myself how I managed to steer Matthew and the shopkeeper to Twink's cage. It was poked away behind a lot of others, right at the back of the place. Then I gave the signal by scratching my left ear vigorously—and Matthew saw that this was one of the birds I wanted. Poor Twink! We bought him for a shilling—Twink, the finest tenor canary that ever sang! But it seems he had had a very sore throat when he came to the shop and had hardly attempted to peep the whole time he was there. So, of course, the proprietor had no idea of his value. Such is fame! Such is life!"

By this time John Dolittle had uncovered Twink's cage and placed it on the caravan table. The bird within, a bright lemon yellow canary, slightly larger than Pippinella, looked very dejected and poorly. When he opened his bill to speak to the Doctor, instead of the golden voice of which Pippinella had so often spoken, a hoarse whisper was all that came forth.

"I have a terrible cold, Doctor," said he. "That fool of a proprietor kept my cage in a draught and my throat got worse and worse the whole time I was there."

"Oh, well, just wait a minute," said John Dolittle, "till I

get you some of my Canary Cough Mixture. It will relieve your throat almost at once, you will see."

Then the Doctor got out his little black medicine bag and from the bottom of it he produced a small bottle containing

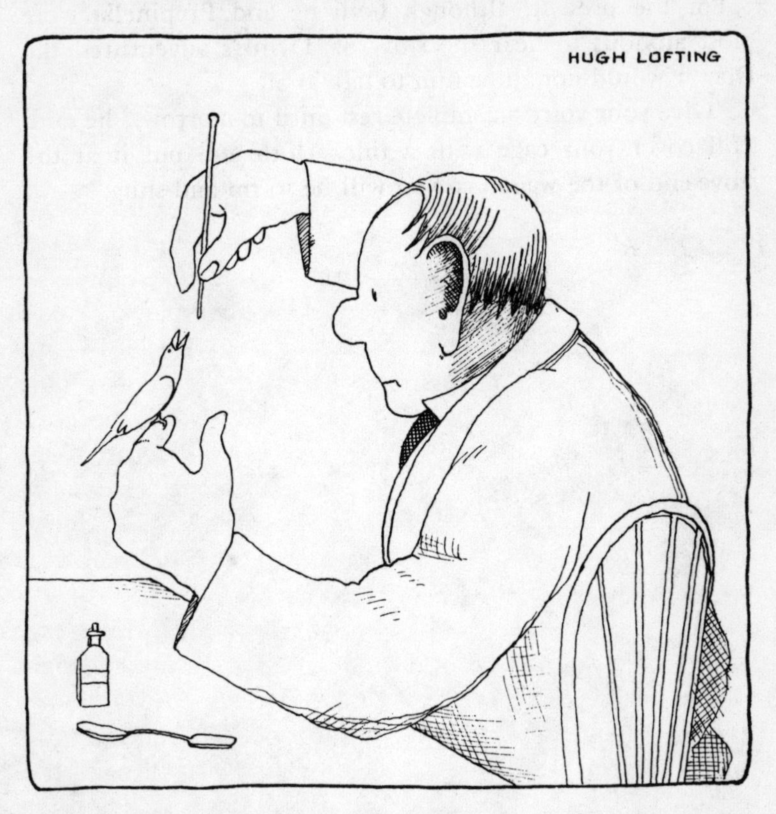

"He let two drops drop into the bird's throat"

a pink liquid. He opened the door of Twink's cage and the bird hopped out on to his hand. Next he took a small quantity of the mixture on a glass rod and when Twink opened his bill he let two drops fall into the bird's throat.

"You will soon feel better now," said the Doctor, closing his bag and putting it away. "Remind me to give you two more drops to-morrow morning. In about twenty-four hours I think I can promise you you will be as well as ever."

For the present, although both he and Pippinella were most anxious to hear the story of Twink's adventures, the Doctor would not allow him to talk at all.

"Give your voice a complete rest until to-morrow," he said. "I'll cover your cage with a thick cloth and put it at the stove-end of the wagon so you will be warm and snug."

PART II

THE FIRST CHAPTER

The Doctor Is Disguised

THE next day, even before he had his second dose of the famous Dolittle Canary Cough Mixture, Twink felt so much better that he was already warbling away softly within his cage when the Doctor got out of bed.

At breakfast he told the story of his adventures—everything since he had been parted from Pippinella right up to his coming to the shop where his clever wife had found him.

"And of all the awful places," he ended, "that shop is surely the worst. I have been in many animal shops in my time, but never in any quite so dreadful, quite so filthy, quite so wretched. There was hardly an animal or a bird in the place that was happy or in full health. The cages were dirty; the food was bad, most of the birds were croupy, the dogs rickety. And do you know, John Dolittle, not half of the birds the man has there were born in cages. Most of his stock is bought from trappers. And, oh, Lord, the sound of those poor thrushes and blackbirds and starlings fluttering, fluttering, fluttering all day long, trying to get out of their cages! Yesterday morning a man brought in a dozen blackbirds for sale which he had caught in the fields. The proprietor bought them all for eighteen pence. This morning two of them were dead, had just battered themselves to death against the wires of the cage trying to get out. It simply made me sick."

This recital greatly saddened the Doctor. His household had often heard him rail against the usual kind of bird shop. But Twink's story of the blackbirds was worse than anything he had ever listened to. For some moments he remained silent. Then he said:

"How are the rest of the blackbirds getting on?"

"No more of them had died when I left," said Twink. "But hardly a one of them was eating any food. Goodness only knows how many will survive. But the blackbirds were not the only ones. Almost every other day some poacher or country lout would bring in a cage full of poor fluttering creatures, linnets, robins, tomtits—everything—which he had caught in traps—all scared out of their wits. Some of them lived and some of them didn't, but all led a wretchedly unhappy existence there."

"Humph," muttered the Doctor. "If it was only the blackbirds, then I could send Matthew to buy them and let them go. But with all those other birds as well it would require a lot of money to do that. It's horrible, horrible! I can't understand how any decent person can inflict such misery on living creatures."

All through supper the Doctor said hardly a word. In spite of his now having all three of his principal singers, he seemed to have forgotten all about the Canary Opera for the present. Twink's description of that bird shop had spoiled the evening for him. Both Jip and Dab-Dab tried several times during the meal to draw him into conversation, but he hardly seemed to hear what they said.

At last when supper was finished he thumped the table with his fist and muttered:

"By Jove, I'll try it!"

"Try what?" asked Jip.

"Listen," said the Doctor, addressing the supper table in general: "do you suppose it would be possible for me to dis-

"Both Jip and Dab-Dab tried several times to draw him into conversation"

guise myself so that no animal or bird could recognize me?"

For a moment after the Doctor's question there was a complete silence in the caravan. Finally Gub-Gub said:

"But, Doctor, what on earth would you disguise yourself

for? I should think you would be no end proud to have all the animals in the world know you."

John Dolittle did not answer.

"It would depend, Doctor, I should imagine," said Jip, "on how well those animals who saw you knew you. For what purpose did you mean to wear the disguise?"

"I want to go to that animal shop," said John Dolittle. "It's worrying me. As you know, I have made it a rule never to go into them, because all the poor creatures clamor at me to buy them. And even if I were rich enough to buy out the entire stock of one shop there would still be all the other animals in the other shops. But this place Twink has told us of seems to be so unusually awful that I thought I'd break my rule and go there."

"And do what?" asked Dab-Dab.

"Let every bird that is not cage-bred go free," said the Doctor.

"Ha ha!" said Jip, getting interested, "I smell an adventure in this. How do you plan to go about it?"

"Well, first of all," said John Dolittle, "I must be disguised so that the creatures in the shop won't recognize me. Then I'll have to get into the place at night or at some time when I shan't be seen or interfered with."

"Good!" said Jip. "When will you go?"

"To-night," said the Doctor firmly. "I shan't be able to sleep till I know that those blackbirds are returned to their freedom. As Twink says, heaven only knows how many will be living by the morning unless something is done."

The Doctor then explained his plan to Matthew: and he, like Jip, entered into the spirit of the idea with great relish—though the Doctor did not seem to care for the way he explained how easily he could pry open the door of the shop to let him in.

"Never mind about those details now, Matthew," said he. "We'll attend to that matter when we come to it. If you accompany me you must understand that we are liable to

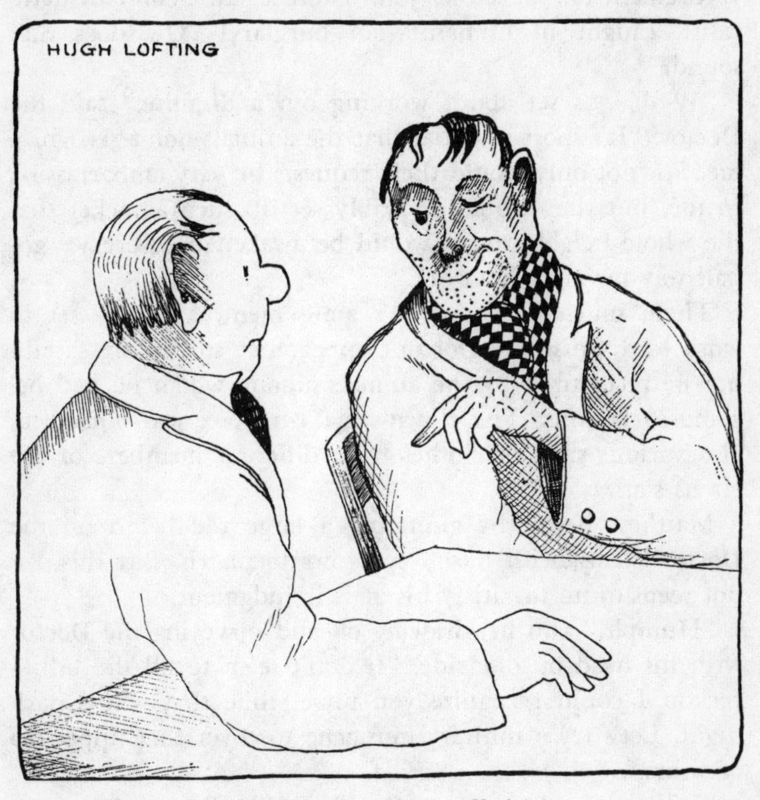

" 'Nifty little party, I calls it!' "

arrest and imprisonment for this. The law will probably call it burglary if we are caught."

"I ain't particular what name the law gives it," said Matthew with a chuckle, "because we ain't going to get caught. Nifty little party, I calls it. That man ain't got no right to keep them poor blackbirds in captivity anyhow. And even if the cops did get us, the magistrate wouldn't be hard on us, you can bet— Look well in the papers, it would—good advertisement for the show: John Dolittle, the heminent naturalist, caught in humanitarian burglary!' 'Ow does that sound?"

"Well, let's set about working out a disguise," said the Doctor. "It's most essential that the animals don't recognize me. For not only would their requests be very embarrassing to me, but they would probably set up such a racket that the whole neighborhood would be awakened before we got half way through our work."

Then, much to Gub-Gub's amusement, Matthew set to work to disguise the Doctor's appearance so that he should not be recognized by the animals among whom he had become so famous. The clown's make-up box was borrowed, also various suits of clothes from different members of the circus staff.

Matthew began by gumming a large red beard on the Doctor's chin, with bushy eyebrows to match. But this did not seem quite to satisfy his artistic judgment.

"Humph!" said he, drawing off and surveying the Doctor with his head on one side. "It don't seem to fill the bill. I reckon I could recognize you myself like that—on a dark night. Let's try a military mustache to cover that upper lip of yours."

"What, more hair on my face?" said the Doctor. "Are you trying to make a monkey of me?"

For answer Matthew gummed a large, flowing red mustache over the Doctor's mouth.

"Good heavens!" said John Dolittle gazing in the mirror. "I look like the butcher in Puddleby. Even if the animals don't recognize me like this, I shall certainly scare them to death."

"Well, you know," said Matthew, "a face like yours isn't

"Surveying the Doctor with his head on one side"

easily disguised. No, I agree. It don't look quite natural. Well, we must try something else."

"Look here, Doctor," said Swizzle, the clown's dog, who was watching the performance with a professional eye: "Why don't you dress up like a woman? The animals in the shop would be much less likely to think it was you. And, besides, as a man you'd never be able to hide that well known figure of yours, even if you succeeded in disguising your face."

"Good idea!" cried John Dolittle. "Listen, Matthew: Swizzle suggests that I get myself up like a woman. Do you think Mrs. Mugg could lend me some things?"

"I'll go and ask her," said the Cats'-Meat-Man. "By Jingo, that's a notion worth two. I don't reckon we could ever do anything with you in trousers and a coat. Wait here a second."

Upon that Matthew ran off and presently came back with not only some of his wife's things but with Theodosia herself.

"I brought the missus along," he said, "because she'll know how the things goes on and can make you look like a real lady. 'Old your face still, Doc, while I get this beard off."

Then while Gub-Gub and the white mouse squealed and tittered with delight, Mrs. Mugg put a skirt and bodice on the Doctor. Next, some sort of wig seemed to be necessary. But the able Matthew made bangs and curls out of the red beard. And, after the back of the Doctor's head had been well covered with a poke bonnet and the bangs were tucked in the front around his temples, he looked like a very nice portly old lady.

"Fine!" cried Too-Too. "No one would ever recognize you, Doctor—not even your sister, Sarah."

"I feel frightfully silly," said the Doctor, tripping over the skirts as he walked toward the mirror.

"Good gracious!" cried Theodosia. "You mustn't walk that way, Doctor. No woman ever walked like that. Take little steps and don't swing your arms so. Now—so—that's more like it. Do you think you better have a veil over your face as well?"

"I do not," said the Doctor. "I'm uncomfortable enough as I am. Besides, I couldn't blow my nose with a veil on."

As soon as the Doctor could walk in a manner which satisfied Mrs. Mugg he set out with Matthew and Jip on his extraordinary expedition.

It was quite a long way from Greenheath to the East End. But even when they finally reached the shop where Twink had been for sale there were lights yet visible in the upper windows. This told them that although the store was closed, the proprietor or some of his household were still up. Across the front window of the shop was a sign reading, *Harris's Song Restorer for Canaries. Four-pence a Bottle.*

"Good heavens!" whispered John Dolittle to his companion. "Sounds like a cure for baldness. I think, Matthew, we had better go somewhere to wait. We can't hang around here. We might arouse suspicion. Let's find a restaurant and have a cup of tea or something. It is now ten o'clock. We will come back in half an hour or so."

So they went off down the street. But it did not seem to be such an easy matter to find a restaurant open as late as this in that part of London. Moreover, the Doctor's skirts were causing him a great deal of inconvenience in walking. Finally when they had reached a very quiet, practically deserted alley Matthew said:

"Listen, Doctor: Suppose you wait here while I go hunting by myself. There must be places around here somewhere. I'm sure I can find one."

"All right," said the Doctor. "But hurry. I have had enough of walking in this get-up."

Then while Matthew went off looking for a restaurant the Doctor hung around the quiet street. Every time any one appeared upon the scene he changed his pace to a brisk walk, so that he wouldn't appear to have no business there. He felt very uncomfortable and unhappy and hoped that Matthew wouldn't be long.

Finally a man and woman came down the street, and al-

"He sprang to his feet and ran"

though the Doctor promptly broke into a smart and business-like gait he noticed that the couple were watching him for some reason or other. Presently he felt that his skirt was slipping off; and at his wit's end what to do, he sat down on a doorstep and tried to look as though he were a beggar-woman taking a rest.

Soon, to his horror, he saw out of the corner of his eye that the couple at the far end of the street were approaching him, apparently with the intention of speaking to him.

As they came near he kept his gaze upon the ground and sat as huddled up as he could so that his face would not be seen. A few minutes later he realized that they had stopped before him.

"Dear, dear!" said the man's voice. "This, wife, is the kind of case one frequently meets with in the slums."

"Poor creature!" said the voice of the woman. "Listen: why do you sit here at this time of night?"

The huddled figure on the doorstep gave no answer.

"Have you no home to go to?" the woman asked.

The Doctor, afraid to remain silent any longer, looked up —and gazed into the faces of *his sister Sarah and her husband!*

Then, grasping his slipping skirt in both hands, he sprang to his feet and ran for all he was worth down the empty street. Sarah gave one shriek and fainted into her husband's arm.

At the first turning the Doctor ran into Matthew.

"What the matter?" asked the Cats'-Meat-Man.

"It's Sarah!" gasped John Dolittle. "And my beastly skirt is slipping off. Let's hide—quick!"

THE SECOND CHAPTER

The Release of the Blackbirds

"SARAH!" said the astonished Matthew as he and the Doctor doubled around the corner and hastened away. "My, that woman has a real gift for turnin' up when she ain't wanted! But 'ow comes she to be 'ere? I thought she was married to the Rev. Dangle, up to Wendlemere."

"Dingle, Matthew, *Dingle*," the Doctor corrected as he panted along the pavement. "Yes, he's one of the canons of the Cathedral there. But I suppose they are on a visit to London. Seem to be out on a slumming expedition, or something of the kind. It's just my luck that they should run into me. Are they following us, Matthew?"

"No," said the Cats'-Meat-Man, looking back. "I don't see no one."

"Well, I must get somewhere where I can fasten up this wretched skirt," gasped the Doctor. "Can't you find me a dark passage or a doorway or something?"

A little further on they came to the vaulted entrance to a stable-yard which seemed to offer the sort of seclusion that they wanted. Making sure that no one should see them go in, they retired into its welcome darkness, and the Doctor did his best to recover his breath while Matthew fastened up his skirt securely. But without any light to see by he got the hem of it so high that the ends of the Doctor's trousers were visible beneath it. This was not discovered till they had ventured out upon the street again. So once more they had to retire to their makeshift dressing room and put the gown right.

"I found a restaurant, Doctor," said Matthew. "Shall we go on there now?"

"No," said the Doctor, getting out his watch with great difficulty from under Theodosia's bodice. "It has gone eleven. And I am sort of anxious about Sarah. I think we had better go back to the shop and leave this neighborhood alone."

So they set off in the direction of the shop. And after about five minutes of walking they came in sight of it, only to find that, although no light now showed in the windows, there was a policeman standing beneath the lamp-post opposite.

"Luck seems to be against us to-night, Matthew," said the Doctor, as they drew back around the corner. "Of all the places on his beat, of course, the policeman *would* have to stop there!"

" 'Ow would it be, Doctor," said Matthew, "if I went up behind him from the other end of the street and tapped him on the head with a screwdriver?"

"Good heavens, no!" whispered the Doctor. "Besides, where would you get a screwdriver from?"

"I got one in my pocket," said Matthew.

"What for?" asked John Dolittle.

"To pry the door open with," said the Cats'-Meat-Man. "I always carry a screwdriver at night—in self-defense, as you might say. But it comes in handy for all sorts of things. Some folks carries a cane or an umbrella; I always carry a screwdriver."

"Well, don't use it on the police," said the Doctor. "That fellow will probably move away if we wait a little while. They have to go around their whole beat every so often. What's the back of the shop like?"

"It opens on to a small yard," said the Cats'-Meat-Man. "But there's no way of getting to it from the street. We'll have to tackle the house from the front."

Then followed a wearisome fifteen minutes while the two, with their noses around the corner, watched the figure of the policeman near the shop.

"At last, with a yawn, the constable stretched his arms above his head"

At last with a yawn the constable stretched his arms above his head and moved off.

"Now's our chance," whispered Matthew, getting out his screwdriver. "Come along."

"And don't forget, Doctor," Matthew added, as they walked down the street toward the shop, "if we get interfered with, begin by talking like a lady, see? We may be able to pass it off as though we found the door open and was just going to tell the owner. But if they get nasty and it looks as though we were going to be grabbed chuck your skirt away and run for it. Here we are. Now you keep an eye open both ways, up and down the street, while I hinvestigate this lock."

"Try not to break the latch or spoil the door, Matthew," said the Doctor. "We mustn't do the man's property any damage. All we want is to liberate those birds."

"Trust me, Doc," chuckled Matthew, setting to work. "I could open this lock with me eyes shut and no one know that I bin near it. There you are! Walk right in and make yourself at home. It's a shame to take the money."

The Doctor turned from watching the street and, to his astonishment, found the door already open. Matthew was putting his mysterious screwdriver back in his pocket as he bowed upon the threshold.

"Good heavens! That was quick work," said John Dolittle, stepping within.

"Sh!" whispered Matthew as he closed the door noiselessly behind him. "This is where the hartistic work begins. Tie these here dusters over your boots while I get the bull's-eye lit— No, I reckon we can see enough with the street lamp 'cross the way. But watch how you tread, for the love of Henry!"

"Get the window into the yard open," said the Doctor. "And as I hand the cages over to you open them and let the birds go into the open air—*Phew!* Isn't the horrible place stuffy?"

John Dolittle's eyes were now beginning to get used to what dim light filtered into the shop from the lamp-post on the other side of the street. It was not a large place, but it was packed and crowded with cages and pens from the ceiling to the floor. In what clear space there was in the middle of

" 'Keep an eye open up and down the street' "

the room stood a line of more cages on stands and little tables covered with dirty cloths of different colors.

The Doctor had been careful to learn from Twink the exact position in the shop of the unfortunate blackbirds' cages and of some of the others that contained starlings and thrushes lately caught and brought here. And as soon as he had his big boots swathed in the dusters that Matthew had provided, he made his way cautiously through piles of pens and boxes till he reached the cages where the blackbirds were kept.

These were pretty large and there were quite a number of them, not more than two or three birds being kept in each. By this time Matthew had the window at the back end of the shop open and a grateful draught of cool air blew into the stuffy room. One by one the cages were handed across to the Cats'-Meat-Man, who deftly opened the little doors and shook the astonished birds out into the night. Of just what was happening they had no idea. But they did not linger to find out. Thanking their lucky stars, they soared upward out of the grimy back yard and winged their way across the chimney-pots of London for freedom and the open country.

"Are they all alive, Matthew?" asked John Dolittle as he handed up the last blackbird cage.

"Not a dead one, so far," whispered Matthew.

"Good," said the Doctor. "I'm glad we were in time to save what was left of the poor fellows. Now we'll set to work on the starlings and thrushes."

John Dolittle next proceeded to find the cages of the other birds who had been caught in the wild and brought here for sale. This was not so easy, because many of the cages were covered; and in the poor light it was hard enough to tell even what kind of bird they contained, let alone whether their occupants were from the country or not.

Moreover, the Doctor had to work with the greatest care lest he wake up other birds in the shop and cause some outcry which might bring down the proprietor or his family.

Most luckily the dogs here were very few—only one retriever and a mongrel bulldog. Both had so far remained fast asleep in their little pens beneath a stack of boxes on the left-hand side of the room.

After a good deal of climbing and searching John Dolittle found a large cage full of starlings and managed to get them across to Matthew and so into the open air without mishap—though one or two of them did chirp a little and were answered sleepily by some bird in the other corner of the room. Two big wooden cages of thrushes were also discovered and emptied in like manner.

Every once in a while Matthew would make the Doctor stop and remain motionless while he listened for any sound or sign from upstairs. But on each occasion the Cats'-Meat-Man was convinced that as yet the proprietor and his household were sleeping soundly, quite unconscious of what was going on below them.

Even after he had set free all the blackbirds, starlings and thrushes he could find, the Doctor still went on wandering quietly around the room peering under cage cloths to see if he could discover any other poor unfortunates who were pining for their native fields and woods.

While he was engaged in this he heard two parrots, whose cages were covered, talking in low tones to one another at the far end of the shop.

"Listen," said one. "There seems to be something moving round the room. Don't you hear it?"

"Yes," said the other. "I thought I heard something, too. It must have been that which woke me up. What is it? Has some animal got out?"

"I've no idea," said the first bird. "I hope it isn't one of the cats. There are two gray Manx cats in the pen near the door. If either of them got loose none of us would be safe."

The Doctor, listening intently, made a sign to Matthew to keep still.

"I'll bet that's what it is," the other parrot answered. "Only cats could make so little noise. And the two of them are loose, I'm sure, because a minute ago I distinctly heard a

"Two cages of thrushes were emptied in like manner"

sound in two corners of the room at once. What'll we do about it?"

"Better screech for the proprietor," said the other. "Because—"

"No, no," whispered the Doctor quickly in parrot language. "Don't do that. You'll—"

"Good gracious!" said the first parrot. "Did you hear that? Some one talking in parrot language! But it's no parrot. The accent isn't right. Queer!"

By this time the Doctor realized that the parrots' conversation had disturbed other bird sleepers round the room. For from all sides came the gentle rustling of wings and scratching of perches. He hastily signaled to Matthew that he wanted to leave in a hurry. But just at that moment he was overcome with a desire to sneeze. He suppressed it as best he could, but the noise he made, muffled though it was, could not be mistaken for anything else.

"For goodness' sake!" said one of the parrots. "Why, there's a man in the shop somewhere!"

As the Doctor once more made frantic signs to Matthew that he wanted to get out and away, he wondered whether he should take the parrots into his confidence or try to get out before the birds guessed who he was. As the second parrot did not answer immediately, but still seemed to be puzzling over the problem, he decided to try and beat a retreat while the coast was clear.

But the parrots within the covered cages did not take very long to realize that there *was* only one being in the world who could sneeze like a man and talk like a bird—even if he did it with an accent. And suddenly one of them cried out loud:

"Why, it must be John Dolittle!"

"Sh!" said the Doctor.

But he was too late. The second parrot was too overjoyed and excited to pay any attention.

"Why, of course!" he squawked. "It's the Doctor. Birds, wake up! The Doctor's come! Wake up! Wake up!"

In a moment every bird in the place was cackling and chattering and screeching and whistling away at the top of its voice. The Doctor made a jump for the door. But in the half light he did not notice a box on the floor. He stumbled over it, came down with a crash and brought a pile of empty cages clattering on top of his head.

"Look out!" hissed Matthew. "I hear steps overhead. We've woken the whole house. Let's get out!"

"I can't get out," said the Doctor. "Take some of these cages off my chest."

By this time, of course, the two dogs were awake and yelping away for all they were worth. As Matthew struggled to get the Doctor's legs and skirt disentangled from the pile of fallen cages he saw the glimmer of a light up the stairs which led to the room above.

"Is it really you, Doctor?" yelped the bulldog from his pen.

"Of course it is," snapped John Dolittle, still floundering around on his back, trying to get up. "For heaven's sake, keep quiet about it! I've got to get away. I'll be put in prison for this if I'm caught."

"Undo the latches of our pens, then," said the bulldog, "and *we'll* see you get away all right."

"Never mind me. Let the dogs out, Matthew," said the Doctor—"quick!"

Matthew, experienced adventurer though he was, never stopped to argue when the Doctor spoke in a voice like that. In spite of the fact that the proprietor was already standing at the head of the stairs with a poker in his hand, he delayed his flight long enough to let the bulldog and the retriever out of their pens.

"Police! Police! Murder! Thieves!" yelled the proprietor, rushing down the stairs.

"Get out, Matthew! I'll manage," shouted the Doctor.

He had not yet quite regained his feet and the proprietor was halfway down the stairs, but again Matthew did not question the Doctor's orders. Leaving his manager to his

" 'You ought to be ashamed of yourself!' "

own devices, he leaped through the front door and disappeared.

Raising his poker above his head, the shopkeeper prepared to jump the last three steps on to his victim's prostrate body. But suddenly two snarling dogs appeared at the foot of the staircase on either side of John Dolittle and dared him to come down. Stiffly the Doctor rose to his feet and arranged his skirt becomingly over his trousers, which had been very evident beneath.

"You ought to be ashamed of yourself," he said severely to the proprietor, as one of his false bangs fluttered from his bonnet to the floor—"for keeping such a disgusting animal shop. If he beats you for this," he added to the dogs, "I'll give you a home. My circus is on Greenheath."

Then he strode out through the open door and set off on the run to follow Matthew, whose figure could just be seen beckoning to him at the corner of the street.

THE THIRD CHAPTER

The Doctor's Return

ALTHOUGH the two dogs successfully kept the proprietor of the animal shop (his name was Harris) from following the Doctor and Matthew, they could not prevent his running upstairs and opening his bedroom window. From there he continued to bawl lustily for the police, till the constable, whom John Dolittle had seen earlier in the evening, appeared upon the scene.

Stuttering with rage, Mr. Harris told him that his store had been broken into and gave him a very minute description of the Doctor. Matthew he had not seen so clearly, but he described him too as best he could.

Then the constable blew a long blast upon his whistle and very soon he was joined by two other members of the police force. Following the direction pointed out by the excited proprietor, they set out on the run to overtake the culprits.

But in the meantime the experienced and intelligent Cats'-Meat-Man was piloting the Doctor through a regular maze of narrow back streets that bordered the river. Presently he stopped and listened.

"I think we've shook 'em for the present, Doctor," said he. "Now let's dodge behind this old warehouse and get them togs off you."

Within the shadow of the shed the Doctor was then divested of his skirt, bodice and bonnet.

"Theodosia will 'ave to do without them," said the Cats'-

Meat-Man as he threw them into the swirling river. "They're no great loss. Next thing, Doc, we must separate and go back to Greenheath different ways. Would never do if the cops

"He continued to bawl lustily for the police"

saw us together—even though you have changed your clothes."

"But what shall I wear for a hat?" asked John Dolittle, watching Theodosia's bonnet drift downstream. "I can't be seen bareheaded."

"I thought of that," said Matthew, bringing a cap out of his pocket. "I brought a spare head-piece along with me, see? With this—and your coat and trousers underneath—I reckoned I could transform you at a moment's notice if we got into a tight place. There's nothing like being prepared."

"Good heavens, it's awfully small!" said the Doctor, trying it on.

"Never mind. Stick it on the back of your 'ead," said Matthew—"So. Now you look all right. And remember, after I've left you, if any cop comes up and asks you questions, remember, you're a greengrocer's assistant on his way to Covent Garden Market. There ain't many jobs, you see, what gets a man up so early as this. So remember: *a greengrocer's assistant.*"

"Do I really look like one?" asked the Doctor, trying to keep the small cap from slipping off the back of his head.

"Well—you'll do," said Matthew. "Turn your coat collar up and don't talk too grammatic—no fancy lingo—see?"

"Very good," said the Doctor. "Which way are you going home?"

"I'm going back through Wapping," said Matthew. "You find some other way and use back streets all you can. I heard that big-footed cop a' blowing on 'is whistle like mad a while ago. So he's likely got all the Metropolitan Police Force on the lookout for us. So long, Doc, I'll see you at breakfast."

Then Matthew disappeared around a corner and melted away into the night, while the Doctor, after looking vaguely about him for a moment, chose a narrow passage which seemed to run parallel with the river in a southeasterly direction.

"I'm a greengrocer's assistant," he muttered as he started off.

He had not gone very far along the passage before a voice suddenly hailed him from behind.

"Hi! What are you doing down there?"

He turned around and saw a policeman, with bull's-eye lantern shining at his belt, not more than ten paces away. To run for it seemed hopeless. He retraced his steps.

"Pardon," said he to the constable. "Were you addressing me?"

"I was," said the policeman. "I want to know what you're doing nosing around the backs of houses at this time of night?"

"I was going to Covent Garden," said the Doctor. "I'm a gardener's assistant."

The policeman turned the light of his lantern on the Doctor's figure and slowly scanned him from his ill-fitting cap to his large boots.

"Covent Garden ain't down that way," said he. "And if you had anything to do with it you'd know better where it was. Come on. Out with it. What's your game?"

There was an uncomfortable silence, during which the Doctor noticed over the constable's shoulder that more people were approaching from the end of the passage.

In the meantime Matthew, experienced in the ways of cities and city police, was gradually approaching Greenheath by roundabout and quiet streets. The gray light of the dawn was just beginning to show when he clambered over the gate of the circus enclosure. In his own caravan he found his wife still sitting up for him, for she was anxious about the results of the expedition.

"It is all right, Theodosia," said he. "The Doctor ought to be along any minute now. I'll just lie down and take forty winks, and then I must get around to my jobs. Wake me up if the Doc gets here before breakfast time."

Breakfast time came, but the Doctor did not. And when, halfway through the morning, he still had not appeared, Matthew and his wife began to get uneasy.

However, about eleven o'clock, just when the Cats'-Meat-Man was preparing to go out and hunt for him he turned up, looking very tired, disheveled and soaking wet.

"Why, Doctor," said Matthew, "what happened to you? I expected you'd be here pretty near as soon as I was."

"A stupid policeman stopped me," said the Doctor, "and asked me a whole lot of questions. He had evidently been warned to be on the look-out for us, but my clothes didn't fit the description, and if I had only been able to answer his questions properly he would have let me go, I'm sure. However, right in the middle of his cross-examination that wretched Harris, the proprietor, appeared on the scene—with another policeman. So—er—I had to employ other methods to get away."

"Well," said the other, "what did you do?"

"I'm afraid I had to use violence," said the Doctor, looking embarrassed and ashamed. "You see, the stupid policeman, when the others came down the passage, got on the far side of me, so I couldn't get out of it at either end. I knew Harris would recognize me, even without my make-up. So—er—I knocked the policeman down with a punch on the jaw, jumped over him and ran for it. At the end of the passage I saw two more constables, cutting me off. There was only the river left. I dived into it, swam under a barge and came up on the other side. They then decided, I imagine, that I had been drowned. Anyway, I heard no further sound of pursuit. And presently I let myself drift downstream a mile or two, crept ashore on the other bank and made my way back here. I was sorry to have to punch the policeman. But what else could I do?"

"Don't apologize, Doc," giggled Matthew, "don't apologize. The only thing I'm regretting is that I wasn't there to see it. But listen: 'ow did you get 'ere like that, drippin' wet, without havin' people looking at you suspicious-like?"

" 'I dived into it' "

"I didn't," said the Doctor. "They followed me in crowds, but no one stopped me."

"Humph!" muttered Matthew with a frown. "That don't sound so good. I don't reckon them cops would give a man up for drowned so easy as you think, specially after getting a poke in the face like what you describe. And if you've left a wet trail behind you all the way from Billingsgate 'ere—with the folks talking—I think you'd better pack yourself away for a while—as soon as you've had a change of togs. 'Cause something tells me we can count on a call or two from the police department during the afternoon— Hulloa! What dog's that?"

The Doctor looked out of his wagon and saw the mongrel bulldog of Mr. Harris's establishment trotting toward him across the enclosure accompanied by Jip, Swizzle and Toby.

"Good morning," said John Dolittle when the canine committee had arrived at the steps. "Did that wretched man punish you for your part in last night's adventures?"

"He wanted to," said the bulldog. "But I and the retriever had agreed to put up a fight together and defend one another if he attempted it. He couldn't lick the two of us at once. Then he tried to separate us and was going to call in help. So I gave the retriever—Blackie's his name—the wink, and we ran off together. Harris came after us, hot-foot, bellowing to everybody in the street to stop us because we were running away. I saw we'd stand no chance of getting to you if we stayed together. So, as we pelted down the street, I whispered to Blackie, 'At the next corner you go one way and I'll go the other. We may reach the Doctor's singly but we'll never do it together!'

" 'All right,' says he. And that's what we did. How he fared I have no idea. But the chances are Harris will succeed in following one or the other of us here."

"I see," said the Doctor. "Well, there's no use in borrowing trouble. He hasn't followed you yet. What's your name?"

"Grab," said the bulldog.

"If Harris does come we'll give him a fine welcome," growled Jip, showing his teeth.

"You bet," said Swizzle. "Listen, Doctor: Grab says he wants to stay here with us. He can, can't he?"

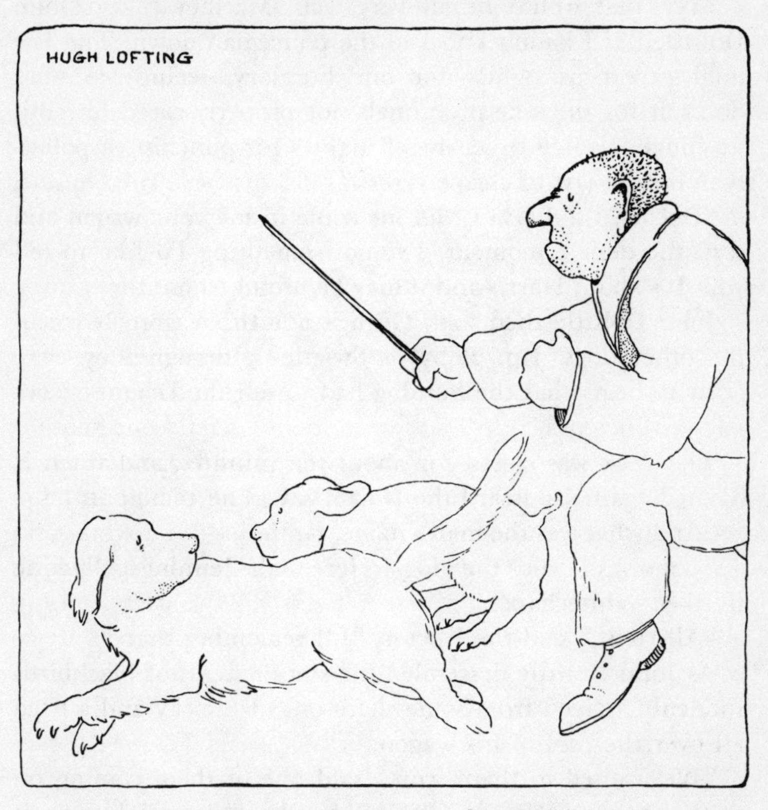

"'Harris came after us, hot-foot'"

"Certainly," said the Doctor—"and Blackie too, if he comes—provided of course we can manage things with Harris. He may not want to sell them to me, just out of spite."

"Don't worry, Doctor," said Matthew, "even if old Harris does come here and we get hauled up for it, they'll only give us a few days in quod. And the advertisement will be fine for the show."

"Yes, that would be all very well, Matthew," said John Dolittle, "if I hadn't knocked the policeman down. The law might treat us lightly for our burglary, seeing we were doing it for the sake of animals not properly cared for. But no judge is going to let me off lightly for punching a policeman on the jaw to escape arrest."

"Doctor," said Grab, "let me come inside your wagon and shut the door a moment. I've got something I'd like to tell you. It's about Harris and it may be useful to you."

John Dolittle then took Grab inside the wagon. None of the other dogs, Jip, Toby or Swizzle—although they were crazy to hear what the bulldog had to tell the Doctor—were allowed in.

The door was closed for about ten minutes, and when it opened again Jip heard the Doctor say as he came out:

"And what was the man's name, Grab?"

"Jennings," said the dog—"Jeremiah Jennings. And he lived in Whitechapel."

"All right," said the Doctor. "I'll remember that."

As John Dolittle descended the steps a flight of blackbirds suddenly arrived from somewhere out of the sky and settled all over the roof of his wagon.

"We wanted to thank you," said one of them coming on to the Doctor's shoulder. "We didn't know last night who it was that had let us out. We were only too glad to get away and did not stop to inquire. But this morning a thrush out in the fields, who is doing some choir work for you, told us that two of his friends had been caught and sold to the same shop. You let them out, too, it seems. And then, of course, we knew who must have done it. So we thought we'd like to come and tell you that we appreciate your kindness very much."

"Don't mention it," said the Doctor—"Good heavens, why here's Blackie—and Harris running after him. This is where the trouble begins."

They had turned, and there, sure enough, was the retriever coming in at the gate at full speed, followed by Harris, the proprietor of the animal store. Jip and Swizzle went for him, snarling like tigers, but the Doctor called them off.

"Leave him alone," said he, "we'll accomplish nothing by that now. Well, Mr. Harris, good morning!"

"Don't you wish me any good mornings," squeaked the little man. "You're a thief. And, now I've run you down, I'm going to get you locked up. My wife can identify you in the police court, as well as I can. She saw you in the shop stealing blackbirds."

"Mind your tongue—calling people thieves around here," said Matthew. "I'm liable to hit you over the crust with a broom handle if you get sassy."

"And you were the other one," said Harris, turning and leveling an accusing finger at the Cats'-Meat-Man. "Good! Now I've got you both, and I'm going off to get the police and land you both in jail. I've got the evidence right here: all them blackbirds on the roof there what you stole and these two dogs that you enticed away. I've caught you red-handed."

"I didn't steal these blackbirds," said the Doctor quietly. "They are at liberty, as they ought to be. You can see I'm not holding them. And, as for the dogs, they came to me of their own free will. They want to stay with me. I am quite willing to buy them from you."

"They're coming back to my shop to be sold to honest

customers," said Harris. "The blackbirds you stole. I found the cages in the shop, with the doors opened, after you'd left. I'm going to get the police."

" 'And you were the other one,' said Harris"

His ugly face purple with fury, the little man turned on his heel and started for the gate.

"Er—just a minute, Mr. Harris," the Doctor called.

"What is it now?" snarled the proprietor, pausing a moment. "I ain't got no time for foolishness. I'm going to get you locked up before to-night."

"I have something I'd like to tell you privately," said the Doctor. "Will you come into my van a moment, please?"

"Anything you've got to say to me you can say in court," snarled Harris, starting for the gate once more.

"I don't think you'd like me to say it in court," John Dolittle called after him. "It's about Jennings—Jeremiah Jennings, of Whitechapel."

The hurrying figure of the proprietor suddenly halted. He turned a scowling face upon the Doctor. Then he came slowly back and walked up the steps into the wagon.

THE FOURTH CHAPTER

Mr. Harris's Past

WIDE-EYED with curiosity over Harris's sudden change of mood, Matthew and the three dogs watched the Doctor follow the man into the wagon and close the door after him.

"I wonder what it is," said Toby in a busybody whisper. "Did you ever see a man suddenly calm down so?"

"The Doctor's got wind of something," said Jip. "That's why Mr. Harris is singing a different tune."

"What was it you told the Doctor, Grab?" asked Swizzle. But the bulldog wouldn't tell them.

"That's my business—and the Doctor's," he said. And that was all they could get out of him.

Meanwhile within the van John Dolittle had offered his guest a chair.

"Let us sit down and talk this over calmly," he said.

"No, I don't want to sit," said Harris in a surly tone. "Tell me, what do you know about Jeremiah Jennings?"

"Well," said the Doctor, getting out his pipe and reaching for the tobacco jar upon the shelf, "I don't believe in raking up a man's past or poking my nose into other people's business. But you have threatened to have me arrested for something which I didn't do. I did not steal those blackbirds, but I did let them go out of your back window because they were wild birds, and you had no business to buy them from the trappers. Incidentally, your animal shop is a disgrace to humanity. But in spite of all that, if I were arrested, in the eyes of the law I should be convicted of burglary and probably sentenced to imprisonment. If, on the other hand, you

"'I wonder what it is,' said Toby"

don't bring any charge against me, nothing will likely be done about it."

"Well?" said Harris as the Doctor paused and struck a match.

"It has been brought to my attention," John Dolittle went

on—"by a very reliable authority—that besides your trade of selling animals you have, or had, another one, that of—er—receiving stolen goods, Mr. Harris. Isn't that so?"

The ugly little man who had been fidgeting around the van sprang forward and thumped the table.

"It is a lie!" he hissed.

"I think not," said the Doctor quietly. "Mr. Jennings, of Whitechapel, who has been in jail for burglary more than once, was one of your best customers. I have enough evidence to put both you and him in prison for a much longer term than you could get me condemned for. I know all about the little—er—'job,' I think you call it, at No. 70 Cavendish Square. I know that you knew the silverware was stolen when it was brought to your cellar for sale. And I know a good deal about Squinty Ted, who had only three fingers on one hand but was remarkably clever at opening safes. And Jeff Bottomley and—"

"Stop!" stuttered Harris. "Where did you get all this? Did Jennings tell you?"

"He did not," said the Doctor. "None of your shady friends told me. I got it from a much more reliable source; and I can prove every word I say. I can even take you to the hole in your cellar wall where you hid the gold candlesticks that came from Lord Weatherby's mansion."

In silent wonder, his ugly face twisted with fear and hate, the little man glared at the Doctor for a moment.

"I believe you're the Devil himself, in disguise," he whispered at last. "Well, what do you mean to do about it?"

The Doctor relit his pipe, which had gone out.

"Nothing at all, Mr. Harris," he said at length—"*Provided* you will agree to a few conditions which I shall lay down."

"Humph!" grunted Harris doubtfully. "And if I don't agree?"

"Then," said the Doctor, "I shall, much against my will, be compelled to hand this information over to the police, to act upon as they see fit."

Mr. Harris thought a moment. Then he jerked his head upward and said:

"Well, what's the conditions?"

"First of all," the Doctor began, "you must agree, of course, not to proceed with this charge you bring against me. After that you must give me your solemn oath that you will receive no more stolen goods or aid in any way burglars or people of similar character. Next, you will have to hand over to me the two dogs, Blackie and Grab—for which, of course, I will pay you. Finally, you must give up the animal shop business altogether. You don't understand it and should never have gone into it."

Harris threw up his hands in despair.

"Do you want to ruin me?" he wailed. "How am I to earn an honest living?"

"Not by selling poor wild birds who have been caught in traps," the Doctor answered. "Nor, certainly, by receiving property which has been obtained dishonestly. You were an iron founder once, I understand—which trade you found very useful in melting down precious metal that was brought to your store. Go back to that work—in the foundries."

Harris made a wry face.

"Yes, I know," added John Dolittle, "it isn't nearly such a comfortable profession as sitting in a shop, selling defenseless animals. But you can likely change it for some other work later."

"Now, look here," began the other in a whine: "you've got a kind face. You wouldn't be 'ard on a poor man like me. I've got children to support, I 'ave. And—"

"No," the Doctor interrupted firmly, bringing his fist

down with a bang upon the table which rattled the china and made Jip outside cock his ears, frowning. "So have the poor birds got children to support—which are just as important to them as yours are to you. My mind is made up.

" 'Do you want to ruin me?' he wailed"

I give you one week to dispose of your stock and close up the animal shop. You're a healthy man. You can earn a living some other way. If the conditions I lay down are not fulfilled within seven days I hand over a written statement to the police, giving your complete record as a receiver of stolen goods. Now, what is your answer?"

Slowly the ugly little man took his hat up off the table. Something in the set of the Doctor's chin told him that further arguments or prayers would be of no avail.

"All right," he said sullenly. "In a week, then."

"Here is half a sovereign," said John Dolittle, taking a coin out of his pocket. "I am paying you five shillings apiece for these two dogs, Grab and Blackie. They will remain here with me. I will come and take a look at your place in a week's time. And as for the receiving game, remember, I have ways of finding out things—those same ways through which I got my other information."

The Doctor held open the door of his van while Harris shuffled down the steps and set off towards the gate.

"Lor' bless me, Doctor!" murmured Matthew, gazing after him. " 'E don't seem so uppish as 'e did. Is 'e on 'is way to the police station again?"

"No," said John Dolittle, patting Grab the bulldog on the head. "I think he'll leave the police alone for good—and the animal business, too."

THE FIFTH CHAPTER

Scenery, Costumes and Orchestra

WHILE Matthew and the five dogs were congratulating John Dolittle over his successful dismissal of Harris the Doctor noticed that the blackbirds were still perched around the roof of his wagon, evidently anxious to tell him something.

"Why," said he, addressing one of the cocks who seemed to be the leader of the party, "I had forgotten all about you during this excitement. I consider it very thoughtful of you to come and visit me. What do you think of my circus?"

"We like it immensely," said the blackbird. "The animals are all so cheerful and clearly enjoying their work. Too-Too has just been telling us about your Bird Opera. And we were wondering if you would like us to help you with it in any way. We are pretty good singers, you know."

"Thank you," said the Doctor. "That's an excellent idea. For, to tell you the truth, I find that I am really in need of another chorus, and your sleek black plumage would look very well against the colored scenery. But time is so short, now. I have promised to have the show in readiness in a week or so. Could you prepare a chorus in so short a time as that?"

"Oh, yes, I think so," said the blackbird.

"Good," said the Doctor. "Then if you'll get about a dozen cocks together I'll read over the score with you and we can decide upon the wording of the lyric for you. By the way, I want a few small birds, as well—very small, to take the parts of a family of youngsters in the nest. What would you recommend?"

"Wrens," said the blackbird. "They're about as small a species as we have in this country. And they're very intelligent. You could teach them anything."

"Do you think you could get me some—four or five?" asked the Doctor.

"Yes, I imagine so," said the blackbird. "I'll go off right away, while it's still light, and see if I can hunt up a few in the woods and farms beyond Highgate."

It was only now, in spite of his having started long before, that the Dotcor was able to give his entire time to the opera. So many interruptions had claimed his attention. But from that morning on until the opening night he worked like a Trojan all day long. Indeed, there were a tremendous number of things that had to be looked after. For, although the opera, when it was finally performed, was put on within the limits of a very tiny stage—not much larger than an ordinary room—yet John Dolittle insisted on the smallest details of the production being made as perfect as possible.

He got the theater owners to supply him with quite a famous artist to paint the scenery for him. And he would spend hours talking over and trying out different arrangements, just to get one little thing right, such as the moonlight glinting on floating seaweed or the evening shadows beneath the hawthorns. In everything he consulted Pippinella's opinion, and if the scenery artist's work did not please her the Doctor would change it, and go on changing it, till she was satisfied.

In the rehearsing, whenever it was possible, he kept the different scenes and passages separate. For instance, with the wrens, when they finally arrived, he practised the nest scene

—where they were to act like a hungry brood being fed by a mother bird—in an old tent quite apart from the other members of the opera cast. He had found that when all the bird actors were together they gossiped and chattered so much that

"He would spend hours trying out different arrangements"

no work was done. And it was his intention to rehearse the different groups by themselves until all were quite at home in their parts, and only to bring them together for the complete performance when they were so sure of what they had to do and sing that there could be no fear of their getting distracted.

The costuming for the opera was not very extensive, but it was difficult. The Doctor decided that only the biggest birds would wear clothes. The pelicans were to be dressed as sailors. The flamingoes would appear as ladies, passengers on a ship. A set of little parasols, light red to match the color of their legs, were specially made by a west-end umbrella firm; and these were carried under their right wings, which were covered with loose-fitting chiffon sleeves.

For dressmaking, of course, the useful Theodosia was called in. Mrs. Mugg, a handy woman with her needle, had always looked after the costumes for the whole circus. For the pelicans she made a set of very natty white sailor suits with little hats to match. And when they were completed, chorus-master Cheapside had to put his troupe through several dress rehearsals of the Sailor's Chorus; because, of course, it was highly necessary that they should learn to walk and behave naturally in clothes they had never worn before.

"Now," said the Doctor to Pippinella, Twink and the other principals one day, "we must take up the question of the orchestral instruments. What do you like best for an accompaniment?"

"Well, of course," Pippinella answered, "it depends somewhat on the song we're singing. Nothing like fiddles or flutes. That's not an accompaniment—it's a competition. The instrument that canaries like best to sing to is the sewing machine. But any quiet, buzzy, humming noise will do."

"Wouldn't you like a piano?" asked the white mouse, who was listening at the other end of the van. "In Puddleby I lived inside the Doctor's piano. I know a good deal about that instrument. I liked best the old one the Doctor had before he went to Africa. That was a German piano, very solidly made—a Steinmetz. It was so warm in winter. After the Steinmetz I like a Wilkinson—an English make. The felt on the hammer is so thick. It's fine for lining the nests for the young ones."

"She made them a set of very natty white sailor suits"

"No," said Pippinella, "a piano's too loud for a canary to sing to."

"Yes, that's true," agreed the white mouse. "I remember one of the Doctor's old patients used to play on the Steinmetz while he was waiting for John Dolittle to come in from the garden. It kept my children awake—until I complained to the Doctor and asked him to lock up the keyboard cover so the patients couldn't disturb us."

"We shall want a razor strop in the orchestra, Doctor," Pippinella went on, "for the duet in the barber's shop aboard ship. That's in the third act, isn't it? It should be an easy instrument to play."

"Very good," said John Dolittle, making a note. "We'll have a razor strop. One of the Pinto Brothers can play on it; and Matthew can perform on the sewing machine. Now, what else?"

"We will need a chain," said the prima donna, "for the Jingling Harness Song—a nice light chain with a clear silvery clink to it. Tie one end to a music stand and get a boy to shake it like this at regular intervals: "*JING—jingle-jing; JING—jingle-jing; JING—jingle-jing.* I begin singing on the fourth *JING*."

"All right," said the Dotcor, making another note. "Anything else? What about the Greenfinch's Love Song?"

"That will be done without accompaniment," said the greenfinch, who was playing the part of the prima donna's faithless lover. "Most of the passages are so soft and whispering that any other noise would completely drown them. But please ask the audience, Doctor, to keep perfect silence during the whole of it, otherwise it will be spoiled entirely."

"I'll attend to that," said John Dolittle, jotting in his notebook again. "I'll have a special request notice printed in the programs. Now, what other instruments will we want?"

"Nothing further besides a cobbler's last and hammer," said Pippinella. "But that instrument will have to be played with a good deal of skill; because we'll use it to keep time in several of the solos—also in the trio at the end of the second act, where my mate goes off with the greenfinch and leaves me mourning on the seashore—also to imitate the raindrops, played very lightly, in the Thrushes' Rain Chorus."

"Very good," said the Doctor, closing his notebook. "Then our orchestra will consist of a sewing machine, a razor strop, a chain and a cobbler's last. I will get them all to-morrow and we will run through the score with music."

THE SIXTH CHAPTER

The Prima Donna Disappears

EVERYTHING went smoothly with the Canary Opera up to within three days of the opening date. Then troubles seemed to descend on Manager Dolittle's head thick and fast. First a serious epidemic of laryngitis and sore throat broke out among the blackbirds; and in spite of the famous cough mixture it ran through the ranks like wild fire. This necessitated not only keeping the rehearsals back (because of course the Doctor could not allow the whole cast on the stage at once for fear the sickness spread to the other birds), but last-minute changes had to be made in the score of the opera when it became clear that the blackbirds would not be cured in time for the first performance.

However, the Doctor packed the birds off into the country (he put the sickness down to using their throats too much in the bad air of the city); and Cheapside suggested that he bring forward a gang of his sparrows to take the blackbirds' place. This was done. And in spite of the fact that the city birds did not have nearly such good voices and nothing like the elegant appearance of the others, the chorus was changed into a sort of comic song and went very well.

As a matter of fact the Doctor had been thinking that he had not quite enough comedy in the opera; and in the end the Sparrows' Chorus in the Fourth Act (where the cheeky birds make fun of the pelican sailors newly arrived from foreign parts) turned out to be one of the most successful numbers in the whole show.

The next trouble was the sudden and mysterious disappearance of the prima donna herself two days before the opera was advertised to open. No one knew what had become of her, and the Doctor was beside himself with anxiety lest harm had befallen his star performer. He had another canary trained as an understudy, it is true. But the posters and all the advertisements laid special stress on the facts that this was the great contralto's first personal appearance in London and that the opera itself was the story of her own life. Jip had disappeared too.

Finally, after the Doctor had put all of Cheapside's city sparrow gangs to hunting for her and called in the help of most of the wild birds around London as well, she was found with Jip over on the east side of London and brought back.

On their return to the circus it came out that on a Saturday morning Pippinella had seen some one in the circus enclosure whom she had thought might be her beloved master, the window-cleaner. Calling on Jip to help her, she had gone off to follow him and the two had tracked the man right across the city, and finally, in the smelly quarters of the docks, Jip had lost the trail and been compelled to give up the hunt. John Dolittle implored his leading lady not to disappear again and said that later, when the opera was in running order, he would see what he could do towards helping her find her friend. At that Pippinella promised not to go off again and rehearsals proceeded.

That same day the Doctor moved his operatic troupe into the city proper, so as to be nearer the theater where the last full dress rehearsals were taking place. A big empty town house was put at his disposal, and for fear of further epidemics and mishaps he kept the different kinds of birds in separate rooms. The pelicans had the drawing room, the canaries the dining room, the flamingoes the big double bedroom on the first floor, the sparrows the kitchen, while Pippinella the star had a room to herself, which had been the late owner's study.

"A crowd of small boys was constantly outside"

The Doctor slept in the basement and Mr. and Mrs. Mugg in the attic.

By this time the birds were all very excited over the nearness of the first performance. And the noise they made running over their songs all day long was so great, even with the front windows closed, that a crowd of small boys was constantly to be seen outside, listening and wondering what was happening within.

As it happened, the week in which the opera was to have its first performance was Christmas week. And when Gub-Gub, Jip and the others of the Doctor's household made a tour through the city they were delighted with the gay holiday appearance of the shops where good things to eat and elegant presents were set out in windows decorated with holly and mistletoe.

They saw many posters and highly colored announcements of various pantomimes and Christmas shows for children. And conspicuous among these were several large advertisements which read:

PIPPINELLA
A CANARY OPERA
In which Madame Coloratura Pippinella, the unique Contralto Canary, will appear for the first time in London, supported by the well known Dolittle Company of Performing Birds.

At THE REGENT'S THEATER
in the Strand
to be followed by the one and only

PUDDLEBY PANTOMIME
(Straight from its smashing success in Manchester!)

Great was their thrill on reading these announcements of their first appearance in the capital. But Jip was almost as much interested in the posters of the other shows. And the Doctor was pestered until he promised to take his whole household to see Dick Whittington (which was being given at the Frivolity Theater) or some other entertainment.

The night before the opera's opening date a grand dinner was arranged at Patti's, a very popular Italian restaurant in the Strand. The theater managers were the hosts and the banquet was especially given for John Dolittle and his staff to celebrate the opera's first appearance in London. The Doctor wore his old dress suit, which he hadn't had on since the days of his regular practice as a physician in Puddleby. Mr. and Mrs. Matthew Mugg came; Hop, the clown; Hercules, the strong man; Henry Crockett, the Punch-and-Judy man; the Pinto Brothers and their wives and Fred, the menagerie keeper.

It was a very jolly meal in spite of the fact that during the second course the Doctor's dress coat (which had become too tight for him) suddenly gave way when he was leaning across the table to converse with one of the guests and split up the back with a loud report. Matthew and Theodosia, who up till then had been uncomfortably overawed by the elegance of the table and the fine dresses of the managers' wives, were by this amusing accident to the Doctor's coat put entirely at their ease, and they enjoyed the rest of the meal with great zest.

At the end, over the port wine, speeches were made by the managers, by the Doctor, by Matthew and by Hercules. The managers said how glad they were to welcome John Dolittle and his opera at their theater.

The Doctor spoke entirely about music and what he hoped to do for musicians and composers by thus bringing forward the musical ideas of the Animal Kingdom.

Matthew Mugg made quite a long speech. In a hired dress

suit his bosom swelled with pride as he spoke of his early ambition to become a showman. He told the managers ("my fellow showmen," he called them) that this was the proudest moment of his life, when he and his partner, the famous John

"Matthew Mugg made quite a long speech"

Dolittle, were welcomed to London to exhibit their greatest creation, the Canary Opera. Such honors, he claimed he had himself foretold long ago, when he helped persuade the great naturalist to go into the show business. Much more he said—

and still more he would have said, if Theodosia hadn't kept twitching his coat tail and telling him in loud whispers not to talk all night.

Hercules made quite a short speech, mostly about the Doctor's system of running a cooperative circus, in which all profits were shared by the staff. This he declared had made him a moderately rich man in a short time, and he hoped soon to retire from the road, for it was his life's ambition to settle down in a seaside cottage with a nice garden where he could grow chrysanthemums and roses.

The Pinto Brothers, when called upon, said they had no speech to make, but offered to give the company a trapeze performance on the restaurant's chandeliers. However, it was feared that these would not be strong enough and the idea was abandoned. Then after some newspaper reporters, who were also guests, had welcomed the Doctor to London in the name of the Press, the party broke up at one o'clock in the morning, and every one went home very happy.

THE SEVENTH CHAPTER

The First Performance of the Canary Opera

DAB-DAB was very worried about the Doctor's health these days. He entered into the excitement of the whole thing with a boyish enthusiasm even greater than that of the performers themselves. He never stopped running about and seemed to be actually in two or three places at once. And the thoughtful old housekeeper shook her head gravely over the possible results.

"I know he's a man of iron," said she, "but for the last three days he has been on the go without rest and has hardly slept at all. Thank goodness the opening night is here at last! For human flesh and blood couldn't stand the pace he's been going much longer."

The Regent's was by no means a small or unimportant theater. There many great actors had produced Shakespeare's plays. It was accustomed to have only the highest and best kind of entertainments; the managers enjoyed a good reputation with the public; and the first nights were well attended by critics from the newspapers. The Regent's could hold nearly two thousand people. And its stage was very large and furnished with all the most up-to-date inventions for lighting, scenery moving, etc.

However, for the Canary Opera the Doctor had had the big stage opening considerably reduced by enormous great curtains of canary yellow silk which cost the managers a large sum of money. The programs were printed on canary yellow paper and the ushers who showed the people to their places were dressed in canary yellow plush uniforms.

Up to the last minute of course John Dolittle was busy behind the scenes looking after the thousand small matters which always get left unattended to up to the opening night, even in the best-run shows. He was being ably assisted by Matthew, Theodosia and Cheapside—who did more swearing on that important night than on any other in the whole of his profane career. The two partners who owned the theater were also doing all they could to help, but it was not much that the Doctor could let them do. Altogether "behind the scenes" was a very busy and excited place.

Jip did not have to dress for his part in the pantomime until much later, and he employed himself during the first part of the evening by running in and out, bringing the Doctor news of the audience and how the tickets were being sold at the box-office.

"Listen, Doctor!" he whispered, appearing for the fourth time and interrupting John Dolittle in the make-up of one of the sailor pelicans: "It's like a mob outside! There are three policemen keeping the people in order round the ticket-office—the line stretches the whole way down the street. And just now a frightfully swell carriage drew up at the door and two ladies and a gentleman got out with diamonds all over them. Maybe one of them was the Queen for all I know—certainly the carriage must have been a duke's at least."

A few minutes later Jip's report was confirmed by one of the managers who had been around to the front on a tour of inspection.

"My friend," he said, grasping the Doctor firmly by the hand, "this is going to be the greatest first night the Regent's ever saw. The seats have all gone, we are selling standing room already and it's still twenty minutes before the curtain goes up."

"What kind of an audience does it look like," asked the Doctor, "intelligent?"

"The best people in town," said the manager. "Come and take a look at them through the peephole. We've specially tried to get the musical folk here, the highbrows and the gentry."

The Doctor, followed by the manager, Jip and Gub-Gub, went to the side of the curtain where there was a little eye-hole through which the actors could see the audience without being visible themselves.

"Well, what do you think of 'em?" asked the manager after the Doctor had looked through a moment.

"Great heavens!" cried the Doctor, his eye still to the hole. "Why, there's Paganini himself!"

"Piggy-ninny!" squeaked Gub-Gub. "Who's he?"

"No—Paganini," the Doctor repeated. "The greatest violinist in the world. That's he in the fifth row, talking to an old gray-haired lady behind him. I've always wanted to meet him. Good! There's one, I think, at least who will understand the music of our show."

The opera's unusual orchestra consisted of Matthew Mugg (sewing machine), George Pinto (razor and strop), Hercules (chain) and a member of the theater's regular orchestra (cobbler's last and hammer). When the musicians trooped in, carrying their strange instruments, a general titter ran through the audience, who did not know quite what to make of it.

And they were still snickering when the conductor (John Dolittle himself) walked to the desk, carrying the usual little white baton. At precisely eight o'clock the Doctor turned to the audience and at once everybody became silent, seeing that some kind of a speech or announcement was going to be made.

The Doctor then told the people in a few words how and

"A little eye-hole through which the actors could see the audience"

why he had devised the Canary Opera. He said that, while the work was intended to amuse and entertain, it must not be taken merely as a farce or burlesque, such as was the Puddleby Pantomime, which would follow the opera. Musically speaking, this production was put forward seriously and its producers felt that it was entitled to the study and consideration of composers and musicians as being the first attempt ever made to bring together the musical ideas of both man and beast. And it was hoped that those serious musicians in the audience would not criticize too early much that was apparently harsh to the ordinary ear, but would wait before giving judgment till the whole four acts had been heard.

Then with a little bow the Doctor turned around and, facing his orchestra, tapped his desk with the stick to command their attention. The audience gave one final fidget (as it always does when the music is about to begin) and settled down to listen in comfortable silence. John Dolittle raised the little white stick, gazed around at his musicians and the overture began.

The orchestra, heard by itself, certainly provided a strange and new kind of music. But, in spite of the odd character of the instruments, the effect was musical without a doubt. The overture was very short, but in a few moments it ran through all the accompaniments and tempos of the main songs of the opera. And the silvery jingle of the chain, the constant droning of the sewing machine, the *rap-a-tap-tap* of the hammer and last and the soft *zip-zipping* of the razor strop provided a mixture of sounds surprisingly pleasing to the ear.

Some of the audience tittered again, but some were clearly not at all inclined to ridicule. One old lady with glasses, in the front row, leaned over and told her neighbor it reminded her of a sleigh ride she had taken in Russia years ago, on the shores of the Black Sea.

"The horses," the Doctor heard her whisper, "galloped right along the snow-laden beach, so near the water that the spray of the surf sometimes showered right over us. That sounds exactly like it—the thumping of the hoofs, the jingle of the harness, the droning of the wind and the whispering of the sea. I remember telling my niece, who was with me at the time, what a wonderful motif it would have made for an

"And the overture began"

opera, if only a composer could have heard it. I'm glad I came to-night."

In five minutes the overture was already dwindling away to a finish, and once more the audience moved expectantly in their seats as they saw from the changing of the lights around the stage that the curtain was about to go up.

The Doctor, as a matter of fact, was now listening with one ear for people going out. Because he had expected that, with anything as new and strange to average audiences as this, there would be some who would go before the overture was finished. But no one left his seat. If the people were not yet enthusiastic, at all events they were interested. And when at last the curtain slid slowly and silently up a general half-suppressed gasp ran through the house at the beauty and novelty of the scene displayed upon the stage.

THE EIGHTH CHAPTER

The Drama of Pippinella's Life

THE entire stage at first appeared to be occupied by an enormous bird cage. But on closer investigation the audience found that one could see through this into a room that lay behind. The setting had been so designed that you got the impression that you were yourself inside the cage, looking outward upon the world. Beyond the large bars, the perches and the water and seed pots, a parlormaid (Mrs. Mugg) was dusting a mantelpiece and bustling silently to and fro over other housework. From a window at the back of the stage a shaft of golden sunlight flooded the cage and the room in which it was supposed to be hanging. High up in the front part of the cage nearest the audience was a nest with the head of a setting mother-bird just visible. And in spite of the enormous size of the cage and the small size of the nest and the birds it did not seem odd or out of proportion at all. You just felt that you yourself had been made very small and put inside a cage with some canaries.

When the parlormaid came forward and pushed a piece of lettuce into the bars of the cage it could be seen that her big figure had been increased by padding and special boots so that she should look the right size when compared with the cage.

There also seemed to be other devices of specially arranged lights and even of magnifying glass screens which made the tiny bird performers clearly visible to the audience at the longest range of the theater.

Presently a second bird who had been down and out of sight near the water pot hopped upon a perch and stood right in the center of the bar of sunlight. Seen thus from the darkened theater, his bright plumage fairly shimmered with

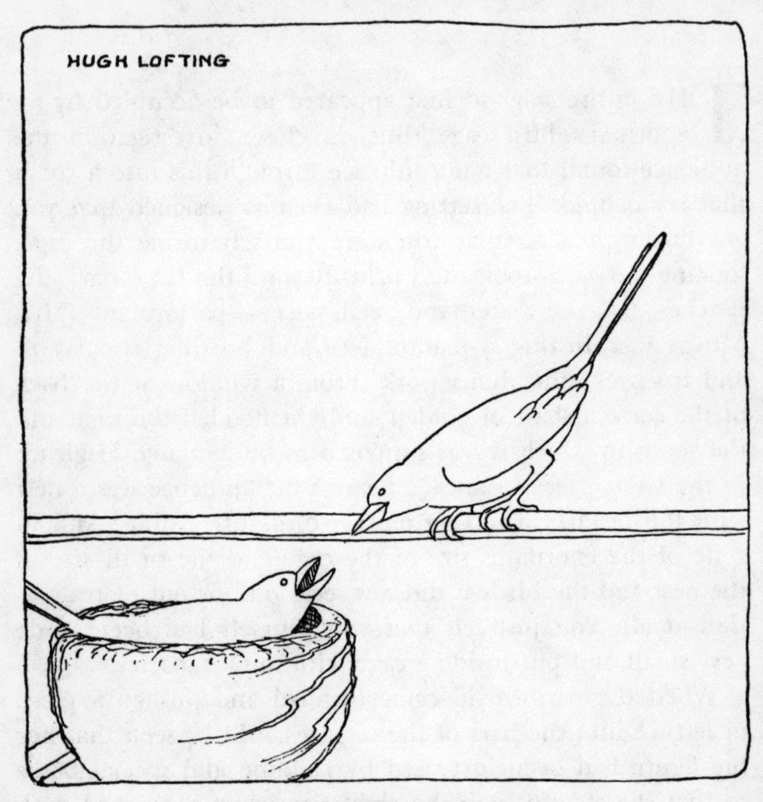

"He fed his wife on the nest"

yellow brilliance. This bird was taking the part of Pippinella's father, for the first act was to represent the great prima donna's childhood and these were supposed to be the nest and the cage where she was born.

After a moment the father-bird went and took some of the lettuce and, carrying it up to the upper perches, fed his wife on the nest. Then the mother-bird got off and the youngsters were fed. The wrens played their parts extremely well. The one who was taking the part of Pippinella as a baby was particularly good. She was the cheeky member of the family and was always reaching up and grabbing the food intended for the others and nearly falling out of the nest half the time.

After the mother-bird had settled down again to sitting on her young ones the father-bird went back to the beam of sunlight and, suddenly facing the audience, sang the first song of the opera. It was only a light little air, mainly about lettuce and sunlight, but it went with a swing and the people were delighted.

Matthew, who was secretly watching the audience while he played his sewing machine, noticed how the few children in the theater influenced the grown-ups. They were not nearly so concerned about their dignity and the importance of their musical opinion as were their elders. They laughed when they felt like laughing and drew in their breath with honest wonder when they were most impressed. They were delighted with the nest and particularly with the naughty member of the brood who was supposed to be Pippinella as a baby. And whenever she poked her nose out from under her mother's wing and tried to imitate her father's singing, the children just gurgled with merriment. Their laughter was catching and soon it had spread through the whole audience.

And, as he silenced his sewing machine at the end of the first solo, Matthew decided that it looked like a good reception for the opera. For the crowd was already in excellent humor, while none of the best voices had been heard yet, none of the choruses had come on and the finest scenery and settings were still to be shown.

At the end of the first act, anxious though he was to hear

"The Doctor left the conductor's desk and hurried behind the scenes"

the comments of the audience, the Doctor had to leave his conductor's desk and hurry behind the scenes. There were a hundred and one things in preparation for the next act that he wanted to attend to personally.

Here he was soon joined by one of the managers, who had been scouting "out front."

"Well," asked the Doctor, "how do they like it?"

"It's hard to say," said the other. "You've got 'em interested. There's no doubt about that. I never saw an audience fidget less or pay better attention. But as for the music, I'm not sure they quite understand it—even the high-brows. I'm not sure I understand it myself. Still, it's going to cause a lot of talk, this, unless I'm a bad guesser. And that's sometimes better than having the audience love you. Anything that promotes discussion and argument gets the public curious and they want to come and hear for themselves. And say, talk about argument! I wish you could step out in front a minute. You never saw so many wrangles going on to the square inch in all your life."

"How did Paganini seem to take it?" asked the Doctor, tying a sailor's hat on a pelican's head.

"He isn't saying much," answered the manager. "But he's thinking awful hard. Sits there wrapped up as always in his usual diabolic calm. But he's interested, too. He'd have left long before this if he hadn't been, you can bet on that. The whole thing will depend on how the critics treat us. So long as we get plenty of space in the papers tomorrow we'll be all right. It's my experience with musical shows that it doesn't matter if the critics say hard things about you so long as they say plenty. Discussion is what we want. And I think we're going to get it."

In the second act Pippinella appeared in person for the first time. The Doctor had especially arranged this in order to save the voice of the prima donna, who was almost continuously on the stage throughout the last three acts. There were several scenes to this act, the curtain being lowered, and the ordinary Regent's Theater orchestra playing, during the intervals between them. The first scene was the inn. Contralto canaries are rare and a voice of Pippinella's quality did not need a musical ear to appreciate it. The first song that the prima donna sang was *Maids, Come Out, the Coach is Here!* It was a great hit. And when she sang the "Jingling Harness" song, to the accompaniment of the Strong Man's chain, some people were so carried away that they stood up in their seats at the end of it and called for an encore. The conductor nodded to his orchestra and the song was repeated.

The second scene was with the Fusiliers. And in this the jolly marching song:

> "I'm the Midget Mascot,
> I'm a feathered Fusilier."

literally brought down the house. The Doctor had arranged to have several of his tent riggers behind the scene tramping to the music. And the audience this time demanded two encores. And even after the second one several people called for more repetitions, until Conductor Dolittle hurried on with the opera, fearing to overtax the leading lady's voice.

Of course, all the incidents in Pippinella's adventurous life could not be crowded into one opera. Yet the main parts were all there: her childhood; her days at the inn; her travels with the soldiers; her underground life in the coal mine (this was a very impressive scene, all dark save a little miner's lamp showing a wooden bird cage on a black wall, with the clinking sound of picks and shovels heard constantly from behind the scenes). Then followed her stay with Aunt Rosie, still another part played by the clever Mrs. Mugg; her restful time at the mill; the storm, beautifully staged, when her cage was blown down; her escape from the cat—for this the Doctor had borrowed the theater cat, who played his part with villainous skill; her meeting with her second husband, the greenfinch, and his heartless desertion of her on the seashore; her flight over the sea and landing on the island.

That brought the opera to the end of the third act. Throughout the whole story it was wonderful to see how John Dolittle had contrived to tell Pippinella's life in stage pictures, so that people who did not understand the language of the actors could yet see clearly what was taking place. This he had often said, while writing the libretto with Pippinella, was the proper way to construct a play, whether musical or unmusical—to have the audience able to see what was happening almost without hearing the words. And in the Canary Opera he certainly succeeded. The people in the theater—and the children, too—never missed anything of what was going on, even the general meaning of the songs was clear to them without their understanding the tongue in which they were sung.

But without any doubt the part of the opera which both musically and poetically most delighted the audience was the scene on the seashore. It was in this that that wonderful mysterious melody, the *Greenfinch's Love Song,* was sung while Pippinella and her husband are hunting for a nesting place. The lights on the stage were dimmed to a pale evening dusk and a little glow of sunset flushed the sky and sea. You could have heard a pin drop in the big crowded theater while the greenfinch warbled and whispered through his trembling serenade, occasionally answered by his mate, Pippinella, from behind the scenes.

The same old lady in the front row got out her handkerchief and began to weep. And when Pippinella, on finding herself deserted by her faithless lover, took wing across the sea to foreign shores she broke down altogether and sobbed loudly.

The scenes aboard the passenger boat, however, with the chorus of boisterous Pelican sailors, who came bellowing into the ship's barber's shop to be shaved to the tune of the razor strop duet, cheered the old lady up no end. And her tears

"The chorus of Pelican sailors"

of sympathy quickly changed to giggles of amusement at the comic antics of the hoarse-voiced bassos.

As soon as the curtain had come down at the end of the last act the Doctor, ignoring the clapping and applause, hurried behind the scenes again to look after the company. But he had hardly reached the dressing room before one of the managers grabbed him by the arm.

"Half the audience wants to kill you and the other half wants to kiss you," said he breathlessly. "Some say you're a humbug and the others call you a genius. But you'll have to make a speech, anyhow. Listen to them yelling for you. And I've brought a special message from Paganini himself. He wants to be allowed to meet you before the pantomime goes on."

THE NINTH CHAPTER

A Triumph and a Success

THAT strange figure of genius, Paganini, had at the time John Dolittle first met him, for years enjoyed a world-wide fame that has never quite been equaled. He had done things with a violin that were never done before nor since. The Doctor, as he had said, had always wanted to meet him ever since he had first heard him play, as a much younger man, in Vienna.

On receiving the message from the manager he called at once for Matthew Mugg, and, giving him a few hasty instructions for the Pantomime, he followed his guide out into the front part of the theater.

Niccolo Paganini was a tall, gaunt man, with a thin face surrounded by straggly hair. Many people who saw him said he reminded them of the Devil himself. He was standing up as the Doctor came down the aisle. And when, after shaking hands with a quaint foreign bow, he invited John Dolittle to take the vacant seat beside him, many people in the audience put their heads together, pointing and whispering. The manager, returning behind the scenes, was very pleased to notice this. Because of course the striking figure of the great violinist was known to every one, and the fact that he had especially sent for the producer of the opera meant that he at all events approved of the strange new music which the Doctor had put before the public. This would make a great impression upon the critics.

"This that you have given us is most interesting, sir," said Paganini gravely. "Do you play any instrument yourself?"

"The flute," the Doctor replied. "But only very amateurishly."

"He was standing up as the Doctor came down the aisle"

"Humph! And have you written much music before this?"

"No," said John Dolittle. "And this, you know, is not my own. This has been composed by the birds themselves. I just did the arranging, the orchestration, such as it is—and that under their direction."

"Indeed?" said Paganini. "But how did you find out what sort of arrangement they wanted?"

"Oh—er—I—er talk bird language," said the Doctor awkwardly. "But I don't, as a rule, speak of that."

"Why?"

"Because people usually laugh at me."

To the Doctor's surprise Paganini showed no sign of doubt or disbelief.

"How absurd of them!" he said quietly, an odd, dreamy light coming into his gleaming eyes. "No one but a fool could listen to your opera without seeing at once that you must have talked with birds and beasts for years to be able so wonderfully to express their ideas in music. What I like about it particularly is that you have not played down to vulgar tastes. And yet it is all so simple, native, natural. You have even included notes in your arias that are so high that the ordinary human ear does not catch them. But my ear is unusual. I could hear them quite distinctly, while most of the audience was asking why the bird kept his mouth still open after the sound had ceased."

"Yes, Pippinella spoke to me of that," said the Doctor. "There is a passage in the *Harness Jingle* and another in the *Greenfinch's Love Song* where the notes go right up beyond the pitch where the human ear can follow."

"You have given us a great treat, sir," said Paganini. "I trust and hope there will be enough really musical folk in London to appreciate what you have done."

By this time the pantomime had begun—to the music of the Regent's ordinary orchestra. And the Doctor, after he had been thanked again by the great musician, made his way back behind the scenes, feeling that he could consider himself repaid for his labors even if no others understood the work he had put before the world.

As a matter of fact, neither the Doctor nor the theater manager had any idea on that first night what a tremendous sensation they had started. For it was one of those successes that begin more or less gently, but grow and grow as time goes on. In the end the Canary Opera, the Doctor's last big achievement as a showman, turned out to be not only the greatest thing in the history of the Dolittle Circus but the outstanding event of the London musical season.

This no one would have suspected at the outset, successful though it was. The Doctor had ordered copies of all the leading London newspapers to be delivered to his wagon on the morning following the first night. In these the opinions varied a great deal, some spoke well of the opera, others badly. But all devoted a good deal of space to it. One or two of them called it the greatest artistic achievement of the century—"a musical revolution." Some couldn't call it enough hard names.

"The most monstrous piece of humbug" (said one paper) "ever set before an intelligent audience was perpetrated last night at the Regent's Theater, when a menagerie owner of the name of Dolittle placed a collection of squawking birds upon the stage and had them accompanied by an *orchestra* of tin-pans, hammers, razor strops and rattles."

By far the majority of newspapers were too cautious to give a downright opinion before they found out how the musical public was going to take it. These called the opera "strange but interesting," "bizarre," "humorous," "quaint," "novel," and so forth. Paganini's presence and apparent approval of the work had made a big difference to many opinions.

The result of all this criticism—good, bad and indifferent —was to make the general public extremely curious. Anything which could call forth such different verdicts must be worth looking into. The second night the Regent's was more crowded even than the first. Furthermore, in those circles of society where music and the other arts are discussed the Canary Opera was for weeks the main topic of conversation. Signor Paganini was called upon by newspaper reporters and asked to publish an opinion. Well known composers who flocked to the Regent's in the first week were also asked to write what they thought for the newspapers. They did—and some continued to condemn it as humbug in no undecided terms. But still the discussion went on, and still the theater was packed tighter and tighter every night.

Soon another problem for yet more talk appeared in the press: who was the mysterious Doctor Dolittle? And could he, as Paganini said he could, really talk with birds in their own tongue? Then the Doctor's empty house in the West End and his circus caravan on Greenheath were stormed all day long by newspaper men, demanding to see him, clamoring to know if this incredible thing could be true.

By no means anxious to be interviewed, sketched and talked at all day long, the Doctor disguised and hid himself and kept out of the way as best he could. But in this he was not very successful. Then Hercules came to his rescue with an idea. He offered to wear the Doctor's high hat and sit in a chair all through the day, with a pillow under his waist-coat, and be sketched and interviewed in his stead.

And that accounts for the many strange portraits of "John Dolittle" that appeared in the London papers about this time—also for many of the extraordinary musical opinions given by him in answer to the questions asked by experts for publication. For, while poor Hercules could wear the Doctor's hat (after padding it to make it fit), he could not wear the Doctor's brains inside it. Indeed, so far as music was concerned, the Strong Man did not know one note from another.

The managers of the Regent's Theater were of course delighted at all this discussion and advertisement. For they

"Hercules is interviewed as John Dolittle"

knew that if it kept up it would be only a question of time before the general public—as well as the "highbrow" or musical—would want to hear the opera.

And, sure enough, by the end of the first week the demand for tickets was so great that they were already seriously considering taking the opera out of their own theater and leasing a still larger one, that could better accommodate the crowds that every night had to be denied admission.

But no one was more pleased with the Doctor's London success than were his own animals. Every evening after the Puddleby Pantomimics had taken their last bow before the curtain they would gather in the dressing rooms to remove their make-up and chat over the night's performance.

"If," Dab-Dab said a dozen times, "we can only keep the Doctor *this* time from spending all the money he is going to make everything will be all right. I've no idea what arrangement he made with the theater owners. But whatever it was, he just can't help, with audiences as big as these, making a small fortune at least."

"Yes, but you must remember," said Jip, tugging his Pierrot suit over his head with his front paws, "that whatever the profits are, the rest of the circus staff has got to share them. There's Hercules, Hop the clown, Toby's boss, the Punch-and-Judy man, the Muggses and the Pinto Brothers. A lot of money doesn't look so big when you've got to split it up among eight."

"I don't care," said Dab-Dab, polishing up the mirror with her ballet skirt, which she had just taken off. "Even so, it must be a large amount. Oh, I do hope the Doctor doesn't go off setting up any more homes for aged horses or sick cats and things, and spending all the money which should take us back to Puddleby."

"Good old Puddleby," murmured Jip, slipping his head into his famous gold collar, which was still his everyday walking-out suit. "My, but it will be good to see the old garden again! And the market place, and the bridge, and the river!"

"And the house," sighed Dab-Dab. "The poor place must be falling to pieces for want of paint."

"It seems a whole lifetime since we were there," grunted

" 'But remember,' said Jip, tugging his Pierrot suit over his head"

Gub-Gub. "Bother this spirit gum! How it does stick! I wish Theodosia would find some other way to keep my wig on. Heigho! I suppose the kitchen garden will be all overgrown and the rhubarb beds smothered in weeds."

"Myself," said Swizzle rather sadly, "I don't share you fellows' rejoicing at the idea of going back. It leaves me and Toby out. When the Doctor leaves the show business I don't suppose we'll see him any more. You can't expect us to be glad he's got rich. It'll be terrible to have him go. He made the circus game into a different life for us. Hop, my man, is a decent fellow in a way. So is old Crockett, Toby's boss. But the world won't seem the same for us after John Dolittle has gone."

"Humph!" said Jip. "I hadn't thought of that before. Well, cheer up! Maybe we can arrange something. The Doctor has got a whole bunch of dogs now, with Grab and Blackie. Perhaps if Hop and Crockett get rich over this they'll let you fellows come with us and the Doctor to Puddleby. It would be a shame to break up the family."

THE TENTH CHAPTER

Animal Advertising

THE success at the theater in the West End affected the business of the circus also. The newspapers in their frequent references to the opera, in their almost daily reports of the notable people who were among the audience last night, spoke of a peculiar and novel kind of circus which this extraordinary man was also running outside the city. And before long, as had happened in other towns, crowds of society folk who ordinarily never went to circuses at all were flocking out to Greenheath. Many important people who had the welfare of animals at heart did a lot of private advertising for the Doctor by talking about the happy condition of his performers, and quite a number of entire schools were taken out there and given peppermints and tea and a perfectly wonderful time by the Doctor, Matthew and Theodosia and the animal hosts and hostesses.

Early in the second week another matter came up, something which had never before arisen out of the Doctor's other successes. This was the subject of animal advertising. The papers and the public were still very much exercised over the question of whether John Dolittle (or the Wizard of Puddleby, as some journalists called him) could or could not communicate with birds and beasts in their own language. That he could do all manner of new things, such as making bird choirs sing in unison, had been proved beyond doubt. Also, there was no question about his having a very mysterious power for getting these animals to perform their strange feats without training, happily and naturally. But whether he could really talk with them was still much debated.

Throughout all this argument and publicity the Doctor

"Society folk came flocking out to Greenheath"

maintained the same modesty that he had always done. He refused to discuss the subject. His work with the animals, he said, was all accomplished by the animals themselves, of their own free will. And whether in that work he actually talked with them or not people must decide themselves from the results.

Well, it began by his receiving quite a number of letters from business firms, asking if he would lend his animals for advertising. The work would be very well paid, the letters said. One manufacturer of bird cages wrote that he would pay twenty guineas a day to have Coloratura Pippinella, the famous Contralto Canary, demonstrate in his shop window in Jermyn Street the superior virtues of his bird cages. She would be wanted for only three hours a day—in other words, the prima donna's time would be paid for at the rate of £7 an hour. All she would be required to do while on duty would be to hop in and out of the cages in the window, to show that even while she was free she preferred this manufacturer's cages to entire liberty, because they were so comfortable and so excellently made.

Then there was another letter which came to the Doctor from a sausage manufacturer. This gentleman offered a similarly large salary to have Gub-Gub do his Pantaloon antics in another shop window. All the pig would be asked to do was to go through his skipping dance out of the Puddleby Pantomime, using a string of the firm's well known Cambridgeshire pork sausages as a skipping rope.

Still another letter, intended to interest the Doctor in animal advertising, contained a request for the Pushmi-Pullyu. This was from a large restaurant who wanted to employ the two-headed animal as a sandwich man. He would only be required, the letter said, to walk through a few of the main streets near the restaurant with signboards draped over his back reading: "Whether You're Coming or Going Take Your Meals at Merriman's Chop House. Table d'Hote Luncheon, One Shilling and Sixpence."

"Good gracious!" cried the Doctor, after he had read a few

of these letters aloud to Matthew Mugg. "What do these people think we are, I'd like to know? I never heard of anything so vulgar or inconsiderate of animals."

" 'I'm not sure, Doctor,' said the Cats'-Meat-Man"

"Well," said the Cats'-Meat-Man, rubbing his chin thoughtfully, "I'm not so sure, Doctor."

"Not sure of what, Matthew?" asked John Dolittle indignantly.

"I ain't so sure you ought to set your face against the idea flat," said Matthew. "Listen, there's this to be said: suppose this feller's bird cages *are* better than anybody else's there's no harm in Pippinella's advertising the fact, is there? It seems to me that if, instead of snubbing these firms outright, you was to offer to do *your* kind of animal advertising a lot of good could be done *and* a lot of money made."

"I don't quite follow you, Matthew," said the Doctor. "The whole idea is very distasteful to me, and I'm sure it would be to my animals. What do you mean, my kind of advertising?"

"Any kind of advertising that would make people think and act more considerate to animals," said the Cats'-Meat-Man, rising from his chair in enthusiastic eloquence. "All your life you've been trying to make folks more thoughtful of the Hanimal Kingdom, 'aven't you? Very well, then, 'ere's your chance. Write to all these here blokes and say, *not* 'Dear Sir: I think you're a low-down hound,' but say 'Dear Sir: I and my animals will be interested in any kind of advertising what spreads the doctrine of the humanitarian treatment of dumb creatures.' "

"Not dumb, Matthew," the Doctor put in. "I've never met a creature yet that was dumb. But I know what you mean. Excuse the interruption. Go on."

"Then," the Cats'-Meat-Man continued, "if they come forward with any scheme that will advertise what *you* want as well as what *they* want, why not go into it?"

"But how about the animals themselves?" asked John Dolittle. "I don't imagine the idea will appeal to them at all."

"Not to old Push," said Matthew. "I reckon he'd die of hembarrassment as a sandwich man. But the others, so long as the work improved conditions for their fellow critters, why, they'd be no end pleased. Take this offer from the cage-makers, for instance: Pippinella wouldn't mind that, I'm sure. And if the cages this man makes are not the kind you approve of, tell him to make the kind you *do* approve of, and that then you'll advertise them for him. See what I mean?"

"Humph!" said the Doctor when the Cats'-Meat-Man finally ended his lecture on the virtues of honest advertisement. "I suppose there may be something in what you say, Matthew. But I fear the opportunities for popularizing the right kind of goods will not be as frequent as those for the wrong."

"Oh, I don't know," said Matthew. "This end of the business ain't hardly begun yet. You haven't even gone through all the letters you got this morning. Didn't you say there was one from the Cattle Committee of the Agricultural Show?"

"Hum—er—yes, I believe there was," said John Dolittle, turning to a pile at his elbow. Out of this, after some rummaging, he took a long, important-looking blue envelope and read the letter it contained to Matthew.

"There you are," said the Cats'-Meat-Man when he had finished. "What could you have better than that? A general hinvitation to cohoperate with the Royal Hagricultural Society for the himprovement of farm animals. Asking you to send in any ideas for the show—for novel exhibits—what occur to you. I calls that a pretty big honor, myself. The Royal Hagricultural Society is no small potatoes. The biggest of its kind in the country. The Queen 'erself is one of the patrons. You see, you're getting a big name now as an animal expert. Well, there's your chance."

"By Jove, Matthew!" said the Doctor, rising, "I believe you're right. Do you remember those drinking cups for cattle that I invented once?"

"You bet I do!" said the Cats'-Meat-Man. "The Dolittle Sanitary Drinking Cups for Cattle—they was designed to prevent the spread of foot-and-mouth disease. But you could never get the farmers interested in 'em. The what-was-good-enough-for father brigade wouldn't let you put 'em on the market. Well, there's something, you see. If you bring them cups forward under the patronage of the Society and gets 'em

"Theodosia entered leading a small, thick man"

demonstrated at the Hagricultural Show you'll find the farmers will treat your ideas very different."

While the Doctor was still thinking over Matthew's words a knock sounded on the door of the room. (This was at the big house in the West End.) And in answer to his "Come in!" Theodosia entered leading a small thick man whose face seemed vaguely familiar to John Dolittle.

"This man wants to see you, Doctor," said Mrs. Mugg.

"My name's Brown," said the visitor. "Last time I saw you you had me hunted out of Blossom's Circus because you didn't approve of the medicines I was selling."

"Oh—ah—yes," said John Dolittle. "Now I remember. Well, it wasn't my fault you were turned out. It was your own. The people wouldn't listen to you as soon as they knew you were selling quack concoctions. What have you come back for?"

"That's the point, Doctor," said the other. "I don't bear no ill feelings for what you did—though I was pretty mad about it at the time. I admit the stuff I sold wasn't much good, but it was harmless anyway. Now I've come back with some good stuff. I have here" (the little man took a bottle with a printed label on it from his pocket) "a real good horse lotion, an embrocation. And I ain't going to ask you to take my word that it's good. I'm going to leave it with you so you can test it and try it out. And maybe if you approve of it, you'll help me sell it. I've had a hard time making ends meet since you downed me on the platform that evening. Just the same, I have faith in this lotion, and I'd like to get your opinion on it."

"Well," said the Doctor in a kindly tone, "you have perseverance anyway. I will certainly have your embrocation analyzed and if I think it contributes anything to animal medicine I will do my best to help you make it known."

THE ELEVENTH CHAPTER

Gub-Gub's Eating Palace

THE official living quarters of the Doctor's animal household had by this time also been moved to the big empty house in the West End. This was done so that the cast of the Puddleby Pantomime would be nearer to the theater. They still, however, spent part of almost every day at the circus. And very grand they felt, traveling back and forth between their city residence and Greenheath, now that John Dolittle could afford to take them in a cab.

The great big five-story house had given them a lot of fun, exploring it from cellar to attic when they first came to live there. Dab-Dab told the Doctor right away that the opera birds living in the kitchen must be put somewhere else if *she* was to do the housekeeping. Some pots and pans had to be bought, too, and a few pieces of furniture.

One of the features of the new house that caught Gub-Gub's immediate attention was a dumb-waiter, a lift intended for carrying the food from the kitchen in the basement to the dining room and the other floors above. He was so fascinated by it that Dab-Dab could scarcely make him stop playing with it for a moment.

"Why, with this thing," said he just as they were about to sit down to supper the first evening they were there, "I could invent a wonderful new game. Look, it runs up to all the floors of the house. D'you know what the new game would be?"

"Hunt-the-onion!" suggested Jip, "or some kind of indoor food sport, I'll bet."

"No, listen," said Gub-Gub. "This is a great idea—one of my best. I would build a palace, an Eating Palace. It would have a great number of floors, and one of these food lifts running from the bottom of the house to the top. All the

"Gub-Gub and the dumb-waiter"

floors above would be dining rooms, but they'd all be different. There would be the Ice Cream Nursery, in the attic. Of course, I'd have to have a good food architect to design the house. Then would come the Pastry Parlor, on, say, the fifth floor. Below that I'd have the Soup Saloon—so the ice creamers above couldn't hear the soup drinkers below. In the Soup Saloon at least a dozen different kinds of soups would be set out on the wide table all day long. On the third floor there would be the Stew Studio—Irish stew, goulashes, curries, etc. Next would come—"

"Oh, stop!" Dab-Dab interrupted. "We know the rest of your silly game: after you had put a different kind of restaurant on each floor, then you'd get into the dumb-waiter and ride up and down, giving yourself a new stomach ache at every floor you stopped. Come and sit down at the table, for pity's sake! You're keeping the supper back."

"Good ideas are thrown away on some people," said Gub-Gub, seating himself sadly before a plate of porridge. "Just the same, that would be a house worth calling a home."

"But where would you sleep, Gub-Gub," asked the Doctor, "if all your floors were dining rooms?"

"In the dumb-waiter, of course," said Gub-Gub, "so as to be ready for breakfast. But you wouldn't have to sleep very much, because by the time you had stopped at all the floors coming down you'd be hungry again, and you'd go back and begin all over from the top."

The pig's delight with this new contrivance and the great possibilities it suggested for the art of eating occupied all his spare time in the new home. He was forever giving himself free rides up and down in the dumb-waiter, deciding where he would have the tomato cupboard and the apple closet and a hundred other new additions to his imaginary Eating Palace. Till one day he got stuck halfway between two floors and missed lunch. Because, although Dab-Dab knew where he was and plainly heard his cries for help, she refused to go to his assistance till after the meal was over, in order to teach him a lesson.

THE TWELFTH CHAPTER

The Animals' Treat

THESE, as you can well imagine, were very thrilling times for the Doctor's household—accustomed though it was to a busy life. Something new seemed to happen every day. And then London at the Christmas holiday season seemed so gay and bustling; just to go through the busy crowded streets was in itself quite an event for any one unused to big cities.

The animals, as I have already mentioned, had frequently pestered John Dolittle to take them to one of the many shows that were going on in town. This he fully intended to do, but had not yet managed it on account of work and also because, of course, most theaters would not allow animals (especially a pig and a duck) in their ordinary seats as audience. But one day, early in the third week of the opera's run, the Doctor for the first time felt that he could afford to give himself a little rest and recreation. So, remembering that he had promised to take his family to a show, he spoke to Matthew about the possibility of getting seats. The willing Mr. Mugg, always on the lookout for anything that would get into the papers and be good advertisement for the circus or the opera, saw a great chance here—though he did not speak of it to the Doctor at the time. All he said was:

"Right you are, Doc. I'll get you seats. What show do the animals want to see?"

"Gub-Gub," said the Doctor, "wants to go to Dick Whittington, the Pantomime at the Frivolity Theater. But the dogs want to see the vaudeville show at the Westminster

Music Hall. There are several trained animal turns on the program there. And Jip and Swizzle and Toby are professionally interested. I think we'd better take them to the music hall—that is, of course, if the management will admit them."

The Doctor had been long accustomed to go through the streets with his strange animal family following at his heels—though, to be sure, of late since he had money he had more frequently taken them by cab for convenience. But he had his doubts about Matthew's getting any theater to consent to his taking them in.

However, the Cats'-Meat-Man did not go to the ordinary booking office to buy his tickets. He put on his best suit of clothes, had a shave and went to call on the proprietor of the Westminster Music Hall. To him he introduced himself in a very lordly manner as John Dolittle's partner. He said that the great composer and impresario was desirous of buying a box for next Wednesday's matinee performance. His party would consist of none other than the original cast of the far-famed Puddleby Pantomime, now playing at the Regent's with the Canary Opera. This party would be made up of three dogs, a pig, a duck, an owl and a white mouse. The appearance of this distinguished theatrical company in the audience would, Matthew reminded the proprietor, be excellent advertisement for the Westminster—particularly since the great John Dolittle, of whom all London was talking, would himself be present. And as the party would occupy a box to themselves the rest of the audience could not possibly object.

The proprietor of the Westminster saw at once that this would indeed be a good thing for his theater. And not only did he consent to let Matthew have the box, but he gave it to him for the Doctor free of charge with the compliments of the management. Then the Cats'-Meat-Man, feeling surer than ever that he was born to be a great showman, proceeded (still in his best suit and manner) to call at most of the newspaper offices in London. There he informed the editors that John Dolittle was taking his company to the Westminster

"The Cats'-Meat-Man went to call on the proprietor of the Westminster Music Hall"

next Wednesday and they must be sure to have a reporter present to write it up for the papers and to make sketches of the famous animals seeing the show.

Great was the rejoicing among the Puddleby Pantomimics

when they learned that their seats had been secured and they were really going to have the theater party for which they had been asking so long. Although he had selected Dick Whittington, Gub-Gub was just as well pleased when the Doctor explained to him that the Westminster's was a variety show—that is, several different turns—animal acts among them. His enthusiasm was not even dampened when Theodosia insisted on washing his face for him (a performance he usually made a terrible fuss about) before he started.

The party was taken there in a cab. At the doors of the theater several newspaper reporters (whom Matthew had arranged for) attempted to interview the Doctor. He objected to this, saying that he was out for a day's holiday. But Matthew persuaded him to stop and say a word to them before he went in.

The ushers, who had been warned, of course, by the manager that this extraordinary theater party was expected, conducted them to their box with all gravity and politeness. On the way there the Doctor remembered that he had meant to bring chocolates for his family, and he was about to send Matthew out to get some, as he felt that no theater party could be considered complete without chocolates. But on entering the box he found that the proprietor had, with rare hospitality, thought of their comfort even in this. On every chair there was an edible souvenir for each of the members of the party. There was a bunch of carrots for Gub-Gub, a piece of cheese for the white mouse, a sardine for Dab-Dab, a small meat pie for Too-Too, a box of chocolates for the Doctor and Matthew and a lamb cutlet for each of the three dogs.

"Very thoughtful!" said the Doctor, biting into a caramel. "Very thoughtful. My! Just look at the audience! The house is quite full."

The animals gazed down over the sea of heads below and around them with a thrill of anticipation. But very soon they noticed that they themselves were the center of con-

"'Dear me, I've been recognized!' said Gub-Gub"

siderable attention. In all parts of the house people were whispering and pointing in their direction.

"Dear me! I've been recognized," said Gub-Gub. "Do you think I ought to get up and bow, Doctor?"

"No," said John Dolittle. "It isn't necessary. And look, the show is just going to begin."

The first turn was a comic tramp with three dogs. This interested Jip, Toby and Swizzle no end. They kept whispering to one another what they thought of the act and at one point in the middle of a ticklish balancing feat, they got so excited that they started barking down directions to the dogs upon the stage. This caused a still greater sensation, because the performing dogs, hearing them bark, looked up and, recognizing John Dolittle, they let their performance go to pieces and started jumping out into the audience to reach the box and greet the Doctor's party. Order was restored after a little, however, and the dogs were caught and taken back to their owner on the stage.

Between the second and the third turn on the program Dab-Dab said to Gub-Gub:

"What are you keeping so still for, with that silly smile on your face all the time?"

"Sh!" said Gub-Gub. "One of the newspaper men is making a sketch of me. He's down in the second row. I want to look my best."

"He's not sketching you," said Jip. "He's making a picture of the Doctor."

At that moment a knock sounded on the door of the box and the proprietor himself appeared to pay his compliments.

"It would be nice," he said, after chatting a moment or two, "if at the end of the next act, when the intermission comes, you and your party would go out for a walk in the promenade. There have been many requests from the audience to get a closer view of you and your company—that is, of course, if you have no objection."

"I shall be very pleased," said Gub-Gub in his grandest manner.

PART III

THE FIRST CHAPTER

The Princess's Dinner Party

WHEN, during the intermission, they appeared in that part of the Westminster Music Hall which was called the promenade, the Doctor's party caused quite a sensation. It was a wide space this, behind the seats on the ground floor, where little tables were set for people to sit and take refreshments. Also, for those of the audience who were tired of sitting down, there was room to stretch the legs and walk about.

As the pig with the dogs and the duck (Too-Too and the white mouse stayed on the Doctor's shoulder to avoid the crush) walked back and forth, they were followed by a train of children who giggled and tittered with delight; while the grown-ups fell back on either side and made a lane for them to walk in, like a street crowd at a royal wedding. Whispers ran in all directions.

"That's John Dolittle himself in the high hat—t'other one, the cross-eyed feller, is his assistant. They say the Doctor talks every animal language there is," said a thick fat man to his wife.

"I don't believe it," answered the woman. "But he's got a kind face."

"It's true, Mother," said a small boy (also very round and

fat) who was holding the woman's hand. "I have a friend at school who was taken to see the Puddleby Pantomime. He said it was the most wonderful show he ever saw. The pig is simply marvelous; the duck dances in a ballet skirt and that dog—the middle one, right behind the Doctor now—he takes the part of a pierrot."

"Yes, Willie, but all that doesn't say the man can talk to 'em in their own language," said the woman. "Wonderful things can be done by a good trainer."

"But my friend *saw* him doing it," said the boy. "In the middle of the show the pig's wig began to slip off and the Doctor called to him out of the wings, something in pig language. Because as soon as he heard it the pig put up his front foot and fixed his wig tight."

Gub-Gub of course put on more airs of greatness than ever here, where all the children—many of whom had seen him performing at his own theater—pointed him out as the great comedian, the first animal actor to interpret the historic part of Pantaloon. Before the bell rang to show that the intermission was over, a school girl had come up and asked the Doctor to sign her autograph album; and numerous other people had interrupted his promenade with congratulations, interviews and what not.

Among these was the man who had written him about advertising the bird cages with Pippinella. He was delighted, he said, at this opportunity of meeting the naturalist-composer in person. He had heard the Canary Opera four times already and was going again the day after to-morrow to take an aunt of his who was coming to visit him from the country. Before leaving he asked had the Doctor decided yet whether he would allow the canary prima donna to advertise for him. John Dolittle said he would first have to consult Pippinella and would let him know in the course of a day or two.

Then another, who seemed to have been half listening to this conversation, came up and introduced himself (with a strong French accent) as M. Jules Poulain, of Paris. He was a manufacturer of perfumes, he said. He had been reading a great deal about Doctor Dolittle of late and also about his now quite famous dog Jip. The stories told in the newspapers (provided for reporters by Matthew, of course) of Jip's marvelous skill in smells, had led M. Poulain to believe that the dog could possibly add a great deal of useful information to the art and science of perfume manufacture. Would the Doctor be willing to let the firm consult him as an expert, and himself act as interpreter for the interview?

Now that the Doctor's household had become theatrical folk, many changes became necessary in the routine of their daily habits. In the old days, when they were only part of a circus, their duties were over regularly by eight o'clock every evening, and the Doctor had always insisted on their being in bed by nine. But now that their work did not end before eleven o'clock at night, early bed was out of the question and late supper had to be instituted. This was something they all greatly enjoyed, when after the show was done they returned to the big city house and Dab-Dab prepared omelettes and salad and cocoa, over which they chatted and made merry.

Another change which, besides the late extra meal, greatly appealed to Gub-Gub was the late hour for getting up in the morning. He had appropriated the dumb-waiter as his bedroom and had it lined with straw cushions. Here the great comedian snored peacefully up to ten or eleven in the morning, at which hour Dab-Dab, preparing breakfast for the company, usually threw a saucepan-lid into the dumb-waiter to wake him up.

Often the hour of retiring was made still later by invitations to supper after the show was over. On one occasion they were all asked to a terribly elegant town house owned by a beautiful Russian princess who delighted in giving unusual parties of any kind. The idea of having the Puddleby Pantomimics, with the far-famed Contralto Canary, as guests appealed to her immensely. Feeling that probably the Princess wanted his company only as freaks to be laughed at, the Doctor had made up his mind to refuse the invitation. But before he had time to send an answer, it was repeated—this time by word of mouth. And, as the messenger who brought it was none other than the great Paganini himself (who had suggested the idea to the Princess in all seriousness), the Doctor was very glad to accept on behalf of his company and himself.

Pippinella and Twink were included in the invitation and the Princess sent a private coach with two footmen to fetch them. The supper was a most gorgeous and splendid affair. All the prominent people in town were there—famous continental opera stars, great composers, writers, painters, sculptors, as well as ambassadors, dukes, earls and a large gathering of lesser nobility.

In spite of this notable assembly, the Puddleby Pantomimics were clearly the guests of honor. Everybody was astonished at the wonderful table manners of the animals and how well they conducted themselves in society. To be sure, one or two little accidents occurred, but nothing of serious importance. For instance, when they first arrived in the gorgeously furnished mansion Jip opened the evening by chasing the Princess's white Angora cat up the chimney—from which she was rescued and had to be taken away to be washed. Then during dinner Gub-Gub (he was seated between a marchioness and the conductor of a symphony orchestra) caused a little sensation when a footman brought around the celery. Instead of taking one stick of celery, Gub-Gub emptied the dish, thinking its entire contents were intended for him. He also somewhat astonished the guests at the dessert by putting his apples and oranges whole into his mouth, instead of peeling them or cutting them.

But otherwise the evening went off extremely well. And after the meal was over Pippinella and Twink delighted the company by singing a duet out of the third act of their opera, accompanied by the Doctor on the sewing machine.

THE SECOND CHAPTER

Jip and the Perfume Manufacturer

AS MATTHEW had prophesied, it turned out that the animals were not only willing to help, but quite keen about the idea when the subject of advertising was put before them.

This John Dolittle did at their after-the-show supper on the same day that he took them to the matinee.

"Why, Doctor," cried the white mouse, "of course the Cats'-Meat-Man is right! You could do no end of good by the proper kind of advertising. And it would be lots of fun for us. Just think of Gub-Gub waltzing with a skipping rope in a pork butcher's window!"

"I will *not* advertise sausages for anybody," said Gub-Gub firmly. "I don't mind working for a skipping rope manufacturer—dancing in a toy shop or something like that. But sausages?—Certainly not!"

"No, I think you're quite right," said John Dolittle.

"But just look at all the other things you could advertise, Doctor," said Jip, "things specially for animals. There's that mange cure you experimented with on me. Remember?—When you were working out a special winter coat thickener for the Eskimo dogs who wrote to you. That was wonderful stuff. My coat grew so fast for two months afterwards it was stiflingly hot in that African climate."

"And how about the famous Dolittle Canary Cough Mixture?" said Twink. "Every old animal shop has a Song Restorer of its own brand, but none as good as yours. Why not give the birds the best?"

"Then there's your system of exercises for flat-footed ducklings," Dab-Dab put in. "If you got every mother duck to raise her children on your system you would do no end of good to the whole race."

"And your hair restorer for mice," squeaked the white mouse, "that you used on me when I got dyed blue by that stupid old rat. That was a tremendously important invention."

"It seems to me," said Swizzle, "that if the Doctor wrote a book on animal surgery and medicine and got it popular with some of those fool vets, a great deal of good would be done. No one realizes except the animals that have been cured by him what wonderful things John Dolittle has added to the science of doctoring."

Well, the discussion went on lively and late over the supper table, all the animals making suggestions to the Doctor on how he could advertise and make known to the world the marvelous inventions and discoveries which he had made in the course of his long and unusual experience with the Animal Kingdom.

The result of it was that before the Doctor packed his family off to bed (about two in the morning) he was made to promise that he would enter the field of advertising and let them help.

"All right," said he, as he took down a candle from the mantelpiece and lit it, "I shall be answering some of the letters from these firms in the morning. And I will see what can be done. At all events, it would do no harm to talk it over with them. If they won't agree to advertise in the way I want them to we can always drop it. To begin with, I think, Jip, I'll see that French perfume manufacturer from Paris. I fancy that your skill in smelling might really add quite a little to science when investigated by technical folk such as M. Poulain and his staff."

The next day the Doctor with Jip went to see M. Poulain of Paris. That well-known manufacturer of perfumes had lately set up in London a branch with experts and workmen brought over from France.

He was delighted when the Doctor appeared with his dog. He had begun to fear that John Dolittle had decided not to act upon the suggestion he had made, nor even to answer his letter. In spite of the visitor's shabby appearance the Frenchman treated him with great courtesy and at once led the way into an inner office. There, after a little polite conversation, he sent for various chemists and foremen and managers.

When these arrived, about five in all, M. Poulain closed the door and addressed the meeting in French, for all his firm were of that nationality, and he had discovered that the Doctor spoke this language with considerable ease.

"Gentlemen," he said, "I take great pleasure in introducing to you Doctor John Dolittle, the eminent scientist. He is the first person in history who has successfully got in touch with the Animal Kingdom. And though many people have cast doubts upon his being able to talk the language of beasts, I for one believe he can. And *if* he can, certainly much will be added to science by his discoveries. He has with him, as you see, a dog. Our business, gentlemen, is the business of scents, of smells. But no man can smell as well as a dog can. Besides, this particular dog is no ordinary one. The gold collar which you see him wearing was presented to him for saving a life at sea, which he accomplished solely by his sense of smell. That is correct, Doctor, is it not?"

"Yes," the Doctor replied (also in French). "I did not know

that the story had been made public. Jip does not like to boast of it. I suppose my assistant, Mr. Mugg, must have given it to the papers. But since it is known it cannot be denied. Yes, it is true."

"Very good," said M. Poulain. "So I think, gentlemen, we can consider ourselves very fortunate in being able to consult this remarkable dog—through the Doctor's interpretation—on the science of scents. And I propose that you now put any questions to him you wish."

Thereupon a sample of perfume was brought forward for the criticism of Jip, the great expert in smells. His first opinion (which was of course translated by the Doctor from dog language into French) somewhat surprised M. Poulain and his staff.

"Why," said Jip, sniffing at a bottle of the firm's very choicest and most expensive perfume, "this smells of cheese!"

"Of *cheese!*" cried the chief chemist springing forward and smelling the bottle himself. "There must be some mistake. No, I don't smell cheese there—only jasmine and honeysuckle. It is a blend we are very proud of."

"I can't help that," said Jip. "I smell the flowers you speak of, of course. But I also detect the aroma of cheese—Camembert at that."

There was great consternation among the experts seated around the board.

"This is terrible," said M. Poulain, wringing his hands. "Just to think that our famous *Reverie d'Amour* should be confused with cheese! I cannot smell it myself, but if the dog can detect the odor, maybe there are others—customers—who can. M. Dalbert, who is the last workman who would have touched this bottle?"

"The man who put the label on," said M. Dalbert, the manager.

"Then send for the chief labeler immediately," said M. Poulain. "And if he or any of his men is in the habit of eating cheese we must change his diet at once. Never, *never* shall it be said that our exquisite *Reverie d'Amour* reminded any one—even distantly—of Camembert!"

Then the manager, M. Dalbert, hurriedly left the council room and presently reappeared with a weedy-looking little Frenchman, who seemed to be scared to death at this sudden summons before the heads of the firm. He was the foreman of the labelers.

"Do you eat cheese?" thundered M. Poulain, leveling a finger at him. "Answer me!"

With tears in his eyes the little man confessed that he was accustomed to take cheese sandwiches for lunch (Camembert cheese), which his wife put up for him in the morning to bring to the factory. But he had always been most careful not to let his lunch get mixed up with the scents. Indeed, he kept his sandwiches in another room, far removed from the bottling and labeling rooms.

"It does not matter," said M. Poulain severely. "It has got into the perfume—or the labels—somehow. You must not bring it to my factory. If you must eat cheese do so when you are on holiday."

So, on the little man's promising solemnly that he would only indulge in Camembert when he was taking his vacation by the seaside, he was dismissed from the council.

Some of the men who had been brought in to question Jip had been inclined at the outset to ridicule the whole performance. Either they did not believe that a mere dog could teach them anything about their business, or they doubted, like most of the public, that the Doctor really was able to understand and translate animal ideas. But after this remarkable detection of the guilty cheese-eating labeler they began to feel that there was something in this of scientific value. Thereupon the consultation of the great expert proceeded in all seriousness, and samples of toilet waters, sachet powders and scented soaps were brought forward for him to test.

But it seemed as though a dog's nose was entirely too subtle, too refined, for a human scent factory. With almost every product put before him (excepting only one or two where the chemical smell was very powerful) Jip's opinion was most unflattering. For he kept detecting some entirely different odor which had been allowed to creep into the prescription.

"I don't think anything of that hair-oil," he said, pushing a bottle aside. "You can smell lard in it a mile off. And as for this soap, the man who made it wore a homespun suit. I can distinctly smell the odor of homespun wool. And the fat used was very rank."

"Try this," said M. Poulain, pushing forward a dainty satin box of face powder. "We always thought this delicious."

"Tobacco," said Jip after one sniff. "It just reeks of tobacco—French cigarettes."

M. Poulain was in despair.

"Well, tell me," he said, after a moment, "what is your own favorite perfume?"

"Roast beef," said Jip. "Why don't you have a roast beef perfume?"

"Oh, la la!" cried the Frenchman, throwing up his hands. "But the ladies do not wish to smell of roast beef!"

"I don't see why not," said Jip. "It's a good healthy smell."

But finally, after Jip had thrown the whole firm into deep gloom by his uncomplimentary remarks about their wares, the Doctor himself turned the proceedings of the meeting in a more profitable direction by remarking:

"It seems to me, gentlemen, that we would derive more benefit from this dog's skill if we allowed him to deliver a lecture on scents in general, or if you consulted him on where, in his own experience, the most delicate and delightful perfumes are to be found."

"You're right," cried M. Poulain, springing to his feet. "Subtlety is what we want in our perfumes, delicacy, finesse, refinement. And that the dog has, heaven knows. But *roast beef,* his own favorite perfume!—Oh, la la!"

"That," said the Doctor, "is only a question of taste. A dog's taste in smells is naturally different from yours. But if I try to explain to him the kind of thing you and your public like I think he will be able to help you."

Upon that, after the Doctor had done a little explaining to Jip, the great expert got up at the head of the table and delivered a lecture on the general subject of smells. That morning was written down in Theodosia's diary of the Dolittle Circus as the first time in history that a dog ever delivered a lecture (even with an interpreter) to a human audience.

It was a great success and was listened to by M. Poulain and his staff (in spite of Jip's former unflattering remarks about their skill) with great attention. The first part of it was devoted to what Jip called "the isolation of smells." For over half an hour the lecturer dwelt on the great difficulty of separating a single smell absolutely by itself—so that you got it quite pure. This, he said, was what was wrong with most of the wares of the firm of Poulain & Co. It was also what was wrong with most humans as good smellers. When a person went down to the seashore, he told them, he would sniff long and loud and say, "How delicious is the smell of the sea!" and let it go at that. While all the time he wasn't smelling the sea *alone* at all. He was smelling a dozen different scents which often come together and which he had grown to call "the salt smell of the sea." The dog professor then gave many learned examples of what he called "group smells"—that is, combinations of different odors frequently met with in ordinary life. Until a person had learned to analyze these, to break them up into their different single scents, he could never hope to be a good smeller.

The latter part of his lecture was given over to describing where his audience could find certain subtle scents, which, from what the Doctor had told him, he supposed would be the kind of thing they were seeking for the perfume market. Many of these were very curious and surprised his hearers not a little. For instance, he told them that down at the bottom of old holes made by a certain kind of field mouse they would find a bed of leaves which were always gathered in the same way and according to the same mixture. These leaves, many of which were aromatic, formed a potpourri more delicate than any he had ever smelled in the finest flower garden. He told them also of a certain swamp bird, a relative of the heron family, that lined its nest with other birds' feathers and with special kinds of moss. This, too, provided a wonderfully delicate, faint bouquet that, in his opinion (and in that particular department of smells), had no equal. Much more he told them—of roots that he had come across, when delving after rats and rabbits, that had wonderful fragrance when bitten or bruised.

All of this information was taken down, when the Doctor had translated it, by M. Poulain's secretary. And that same week the Paris perfumer hired botanists and naturalists to go out in pursuit of the ingredients that the famous expert had described. The result was that a few months later several new perfumes and powders were put on the market whose delicate fragrance wafted the fame of Poulain, of Paris, to every corner of the earth where refined noses were to be found.

THE THIRD CHAPTER

The Ways of Advertisement

THE Doctor, in his usual unbusinesslike way, had made no money arrangements with M. Poulain for Jip's services as an expert in smells. Nevertheless the Frenchman behaved very handsomely. And a check for a considerable sum of money arrived by mail not long after the interview I have just described.

"Humph!" said the Doctor, taking it out of the envelope and noting the amount. "Well, this money should belong to you, Jip, by rights. *I* certainly did not earn it."

"Oh, what would I do with money, Doctor?" said Jip, poking his nose out of the window to sniff for rats in the back yard. "You keep it. You're always in need of money. And besides, don't I owe you a whole lot for the good home you've given me all these years?"

"There you go, Doctor," said Dab-Dab, in a flutter. "Always, when money comes in, you want to disclaim owning it. I never saw any one like you."

"But Dab-Dab," said the Doctor, "we're making so much by the opera now. It isn't as though I were poor any more. Last week I paid off every single debt the circus owed since Blossom left. And still my bank account shows a bigger balance than I ever had in my life."

"No bank balance is too big for you to spend," said Dab-Dab. "My only worry now is to get you back to Puddleby before you're penniless again."

"Just the same," said the Doctor, fingering the check, "I feel I ought to put this money in the bank in Jip's name. Who knows? The day may come when he will need money. Suppose anything should happen to me . . . I wonder if a bank would accept a savings account in a dog's name. A very interesting point. I must look into that."

Jip's performance with Poulain & Co. led a few days later to another offer from a firm of dog-collar makers who had a small but smart shop in St. James Street. This was a company who specialized in all sorts of things for pets. Here society women came to buy silk sweaters for their Pomeranians, porcelain drinking bowls for lap dogs and expensively upholstered sleeping baskets for prize Persian cats. It was a prosperous, well-to-do house. And in the letter they wrote the Doctor they said that as for the way in which the advertising was done, they would gladly leave that entirely to him and his animals.

Jip when he was told about it was not very enthusiastic.

"I know that kind of shop," he said. "They sell all sorts of absurd faddy things for overfed pugs; satin ribbon collars, rubber pacifiers for puppies, shaped like bones, and all sorts of other rubbish. I'd sooner not have anything to do with a shop like that. Why should these pampered pets of society dames have silk sweaters round their silly fat stomachs, when there are hundreds of good dogs—real dogs, even if they are called mongrels—slinking round the East End of London trying to pick up a square meal?"

The Doctor agreed with Jip and the matter was dropped. But Dab-Dab, who had overheard the conversation, took Jip aside later and explained to him that if money could be made for the Doctor in this way it would be a good thing.

"Because, listen, Jip," she said, "the more money we have the surer—and the sooner—we'll get the Doctor back to

Puddleby. And most of it will be spent on deserving animals, in the end, anyway."

"All right," said Jip. "I'll see if I can't arrange something. But I *won't* advertise satin collars or cat's cradles."

Jip then set to work and with the help of the Doctor he got up quite an interesting show, to be given in the window of the dog collar shop. He called in as assistant actors all the other dogs attached to the Dolittle caravan: Swizzle, Toby, Grab and Blackie. They were a well-mixed company. Swizzle was, of course, as usual, the comedy dog, making a joke of everything and everybody. Toby was the self-important, cheeky, small dog who just *has* to boss anything that's going on. Grab looked the picture of a ferocious fighter (though he was really quite a good-natured animal); while Blackie, the retriever, was a good sample of the larger, more serious type of dog.

The little play that they arranged was quite simple; and a good deal of it was not planned ahead, but contrived on the spur of the moment. They just sat or stood around in the window as though it were a sort of dogs' club. And Jip, pretending he was a pedler of wares, brought to them collars and winter coats and dog soap and what not for sale. Toby acted with great skill the part of the fussy sort of spoiled pet who was pleased with nothing. Blackie played the more dignified rôle of the large dog on whom were tried all the big collars and coats and things. Swizzle, as the buffoon, pretended he didn't know how any of the things were worn. He put the collars on inside out, the coats on upside down and generally tripped over, or bungled, everything.

Live animals in a shop window will always attract the attention of the passer-by. But such a gathering as this, of dogs actually doing things, brought a crowd around the shop such as had never been seen before. Policemen came to see what the matter was. And of course their presence brought still more people who thought something serious was taking place. It was quite the best advertisement this shop had ever obtained.

Matthew Mugg, with his usual business cleverness, had insisted on the shopkeepers putting a sign in the window telling the public that these dogs were part of the far-famed *Dolittle Animal Theatrical and Operatic Company.* And of course this was very good advertisement for the Canary Opera, the Puddleby Pantomime and the Circus on Greenheath.

Gub-Gub, who had refused to enter the field of advertisement to help sausage manufacturers, did so later with great success in a West End toy shop. Here, amidst delighted children, who romped around him all day long, the pig comedian jumped a skipping rope, played with mechanical toys and generally amused himself as well as his audience.

Pippinella, too (after she had privately inspected the cages made by the manufacturer who had written for her services), fulfilled her contract. Hers was the very simple task of hopping in and out of cages, showing by her lively and sprightly manner that she approved of their design. But the fame of her wonderful voice had now spread so far that her mere presence in the shop more than repaid the firm who had employed her at the enormously large salary of £7 an hour.

THE FOURTH CHAPTER

The Pocket Horse

BUT by far the most useful venture into advertising undertaken by the Doctor or his household was with the Royal Agricultural Society. This large and important association was accustomed to hold its annual show in one of the biggest halls to be found in London. And not only did it occupy the hall itself, but its sideshows and smaller exhibits spread over into adjoining grounds and other buildings. It was a yearly event of long standing with the public. It ran for a fortnight and it was well patronized, not only by the citizens, but by great numbers of farmers and others interested in agriculture and stock raising who came from every corner of England to attend it.

Each day in the fortnight was devoted to some special department, though of course there was much that was on show continually throughout the two weeks. One day there would be cattle-judging, when prizes were given for the best-bred cows or the fattest sheep. Other days a parade of Shire horses—to which would be sent the finest stallions and champion plowing teams from all over the country—would take place. Still another day would be given over to butter-churning competitions, dairy demonstrations, fancy poultry and so on.

Now as soon as Matthew Mugg had learned that the Doctor had been written to by the Royal Agricultural Society he had, without consulting his manager, made a call upon the Exhibition Committee. The show had not yet opened, but was due to do so the following week. To this committee Matthew had introduced himself in his usual manner, as the partner of the great John Dolittle, M.D. He congratulated the committee on their wisdom and foresight in inviting the renowned naturalist to work with them for the success of the Royal Agricultural Show. Then having once broken loose on his favorite subject he went on for over an hour telling the committee of the astonishing discoveries, inventions and scientific achievements of his partner. Much of it was even more astonishing than anything the Doctor had ever done. But it was convincing enough to greatly impress the Committee.

The result of this little advance expedition of Mr. Mugg's was that the Doctor was called upon two days later by the Chairman and Secretary in person. This, even for the famous Doctor, was no small honor. And he took great pride (he was at Greenheath when they called upon him) in conducting his distinguished guests round his circus.

They were delighted at the wonderful condition of all the animals kept there. Fred took them through his model menagerie and showed them elephant stalls and lions' dens such as they had never seen before.

But it was with the Doctor's circus stables that they were most impressed. These were John Dolittle's pride. The enameled iron hay mangers, the white earthenware drinking troughs, the washable cotton tether ropes and blankets, the ventilators which the horses could open and close themselves, the marvelous cleanliness, the atmosphere of health and good cheer, fairly took their breath away. They asked the Doctor how had he accomplished all this, where had he got these ideas from? He giggled and changed the subject. The Chairman of the Committee insisted, however.

"Come! Don't you want to tell us?" he asked.

"Oh, yes," said the Doctor. "I'm quite willing to tell you. But you probably won't believe me—most people don't. You see, practically every detail and device in these stables has been invented by the horses themselves. These ideas are their own—this is their conception of a well-run stable. I have merely carried out their wishes. I—er—talk their language, you understand."

Well, the outcome of the Chairman's and Secretary's visit to Greenheath was that the Committee was keener than ever to have the Doctor take part in the Agricultural Show. And he was sent a special invitation to visit the hall and grounds two days before they were thrown open to the public, to help with his advice on the livestock housing and to give the Committee the benefit of his ideas in any other departments.

On this trip Matthew, of course, accompanied him. And likewise did Dab-Dab, whose opinions he wanted on the Aylesbury Duck Exhibit. Jip also came, because he was very interested in a special demonstration of trained sheep dogs which the Society was giving, and Gub-Gub, too, joined the party, as he wanted to see the prize potatoes.

They found the premises of the show even greater than they had expected. Some of the stands and tents had not been put up yet and many were still in the course of building. But even so, the extent of the whole affair was enormous. There was something of everything that has, or ever has had, to do with farming: plows, harrows, harvesting machinery, incubators, chicken houses, traveling boxes for eggs, sheep wool clippers, sheep fencing, prize vegetables, tomato hothouses, chemical apparatus for testing soils, magnificent horses of all breeds, from the heavy Shire horses to little Shetland ponies. Altogether, indoors and outdoors, there were literally acres of things to be seen.

Matthew, of course, as usual, was very much on the lookout for opportunities to get the Doctor to show off his knowledge. And as John Dolittle stopped on his way round to admire some Shetland ponies that were being shown the man who was exhibiting them came up and spoke to him.

"I hear, Doctor," said he, "that you have done some quite interesting work in pony breeding yourself."

"Oh, dear!" laughed the Doctor. "Has Matthew been bragging about that, too? Yes, I have done a certain amount—not very much."

"Why!" cried Matthew. "What about that dwarf horse you produced?—Smaller than any of these Shetlands here, wasn't it?"

"Oh, yes," said John Dolittle. "It was certainly that."

"Well, smallness is what we want in Shetlands now," said the breeder. "You can get any prices for good teams if they're well matched and small enough. We'll be glad to learn anything you can teach us."

"The occasion Matthew speaks of," said the Doctor, "was some years ago in Burma, when I was on a voyage. I learned of a special kind of rice which, when cattle were fed on it from birth, kept their size down in a remarkable way. They had a craze there, too, for small animals at that time, and I thought I'd like to see what I could do. After some experimental work I produced a pony which I could put inside my hat. A very intelligent little creature. But I did not repeat the breed at all. I found that he was not happy when brought down to such an unnatural size. Of course, he got petted to death by all the Burmese ladies and had an excellent time that way. But one day a dog mistook him for a rabbit and ran off with my horse in his mouth and I had quite a time overtaking him. I then realized that I would have to restrict his liberty a great deal if he was to be saved from a whole lot of enemies who, were he of ordinary size, wouldn't have dared to attack him. And that didn't seem quite fair. I finally gave him to the King of Siam, who had a special little garden made for him, with a big wall around and a grating on top, to protect him from hawks. But I made up my mind that I would breed no more pocket horses."

The part which the Doctor finally played in the Royal Agricultural Show turned out to be very useful. For one thing, he got the Committee to give him a special stand with half a dozen fine horses and cows to demonstrate his sanitary drinking cups for cattle. These he had invented some years before to prevent the spread of foot-and-mouth disease and other animal complaints. The cups were self-draining contrivances which, when the horse pushed his nose into them, turned the water on and filled themselves automatically. They also drained and rinsed themselves when not in use. The cattle were thus always provided with fresh, clean water.

The horses and cows who gave the demonstration very soon got on to the workings of the Dolittle sanitary cups and throughout the show there was always quite a crowd gathered about their stand watching them.

Then another thing which the Doctor introduced was an animal drug-store. Here were displayed all the medicines and embrocations and soaps and hair restorers and other things which he had found in his long experience as an animal doctor to be good, reliable products. Among them was Brown's horse liniment, which John Dolittle had analyzed and found to be very good. Also his own canary cough mixture (which worked equally well with poultry) and his mange cure and many more.

Further, in this stand he also demonstrated every afternoon in person many discoveries he had made in animal surgery.

Most of the patients who daily flocked to Greenheath were directed by Jip and Dab-Dab to come here instead. And the Doctor set bones and cured all sorts of ailments publicly in dogs and calves and horses so that those who were interested in the science of animal surgery could get the benefit of his experience.

This animal drug-store and surgery was frowned on by many of the vets who came to the show and tried to discredit the Doctor as a quack. But so extraordinary were many of his cures, that finally there was no doubt in the minds of the public about the genuineness of his remarkable skill. Also the fact that he held a degree as a graduated Doctor of Medicine made it difficult for the envious ones to arouse feeling against him. And very soon the veterinary surgeons became convinced themselves that he knew more than they and they were only too glad to learn from him.

During the last week of the show the Doctor's consulting hours became a sort of class in veterinary surgery and medicine. You could hardly see his stand at all for the crowds of vets and students who watched and listened while he set the broken shoulder of a sheep; relieved a limping horse; put gold fillings in cows' teeth and performed delicate and wondrous feats of surgery which had never been seen before.

The Committee said at the end of the show that this year the general admission at the gate had been twice as large as it had ever been before, and they put this extraordinary interest of the public down to the fact that Dr. John Dolittle had taken part in their annual exhibition.

THE FIFTH CHAPTER

The Dolittle Circus Staff Holds a Meeting

THE Canary Opera's run at the Regent's Theater broke all records in successful theatrical productions. Week after week went by and instead of the attendance falling off, the house seemed packed a little tighter—and the crowd turned away a little bigger—every night.

The One Hundredth Performance was celebrated with another dinner at Patti's in the Strand. But on this occasion some of the faces that surrounded the table at the first celebration were absent. The reason for this was that many of the original staff of the Dolittle Circus had made enough money to retire. Hercules had gone to his peaceful cottage and rose garden by the sea; the Pinto Brothers had departed; and so had Henry Crockett, the Punch-and-Judy man.

Furthermore, when the meal was over and speeches came to be made, Hop the Clown arose and told the company that this was his farewell appearance in public. He too, he said, had made enough money to retire and much as he disliked the idea of parting from his excellent friend and manager, John Dolittle (cries of "Hear! hear!"), he had wanted all his life to travel abroad. And now at last he had the means to do it. His dog Swizzle, however, had chosen to remain with the Doctor, so that he did not feel he was altogether deserting the show, since his best friend, the companion of many years of circus life, would remain in the ring to keep his memory green.

At this dinner the dress of the company was noticeably different from that worn at the first. Mrs. Mugg, the Mistress of the Wardrobe and chronicler of the Dolittle Circus, fairly shimmered with jewelry. Matthew, too, when he arose and launched into a long and eloquent speech, was seen to be wearing three enormous diamonds in the bosom of his shirt and another, even larger, in a ring upon his finger. The Doctor himself wore a brand new dress suit made by a Bond Street tailor; but he said it did not feel nearly as comfortable as the one that had split up the back at the dinner given to welcome his operatic company to London.

The Doctor, in his own address, spoke of the disappearance of familiar faces from the dinner table. He said he was glad that his remaining so long in the show business had given opportunities for many to realize their ambitions and to do those things they wanted most to do. Money itself he always regarded as a terrible curse, he said, as it too often prevented people from doing the things they most wished for, instead of helping them. They had now been in London three months. And the budding of Spring on the park trees reminded him of his own home in Puddleby, of his garden and his plans which he had so long neglected. Before another three months had passed, he said, he hoped that he himself would be able to retire from the theatrical world, and that the other members of the company who had made the Canary Opera the success it was would be sufficiently well off to do likewise, if they wished.

The departure of Hop the Clown from the Dolittle Circus was an event of importance. Both Swizzle and he wept on one another's necks when they came to say good-by. Swizzle was torn between two loves. He wanted to stay with Hop, but he also wanted to stay with John Dolittle and his jolly, crazy household. He could not do both. And it was only after Hop had promised, through the Doctor, that he would keep him posted by letter on how he got on, that the circus dog was consoled. Even then all that night he kept Jip and Toby awake, having qualms and scruples about deserting his old master.

"You know," he'd say suddenly when the others were just feeling sure and glad that he'd gone to sleep, "it isn't as if I *had* to let him go alone. They say a dog's place is by his master's side. He was always an awfully kind, considerate sort of boss, was Hop. I feel a terrible pig deserting him after all these years."

"Oh, forget about it and go to sleep!" said Toby irritably. "I don't see what you're blaming yourself for. You worked for your living. You helped his act in the ring—in fact, you were the better clown of the two. The audiences always laughed more at your antics than they did at his. Now he's going out of the show business with plenty of money—which you helped him make. If you prefer to stay with Doctor Dolittle while Hop travels round the world enjoying himself that's your business."

"Yes, I suppose so," said Swizzle thoughtfully. "Still, he was such a decent fellow, was Hop."

"And so was Henry Crockett, my man," said Toby. "But I left him, or rather he left me, in the same way. I wanted to stay with the Doctor, too. I helped him make the Punch-and-Judy show a success. I don't feel guilty. My goodness! After all, our lives are our own even if we are dogs."

Well, finally Swizzle was persuaded that he need not let his conscience bother him about Hop; and Jip and Toby got a little undisturbed sleep somewhere between two in the morning and getting-up time. As it happened, the Doctor himself was wakeful, too, that night, thinking over various problems connected with the opera; and as the dogs' quarters were situated in the passage outside his room he had heard the whole conversation.

What Toby (who, as you will remember, always insisted upon his rights) had said about their lives being their own,

"'Oh, forget about it and go to sleep,' said Toby"

even if they were dogs, set John Dolittle thinking. And the next day, after he had spent more than an hour talking over money matters with Too-Too the accountant, he called a meeting of the circus staff.

For this he went out to Greenheath, taking his household with him. It was an odd gathering. The animals present outnumbered the humans by far. The Doctor himself had not fully realized up to this how the ranks of his performers had thinned. In one of the larger sideshow tents where they were all collected, Manager Dolittle rose and made an address, first in English to the people and then in animal language to the rest.

"The cooperative system," he said, "which we have followed has proved, I think I can say, a great success. But I feel that it is only fair that now when we have a big balance in the bank, due largely to the Canary Opera's reception, the animals who have taken so important a part in all our shows should share in the profits. And in any case it is desirable that all our performers should be provided for when we disband. I therefore propose that the animal members of the staff share with the rest of us on the same basis and that bank accounts should be opened in their own names. I have looked into this matter and find that it is quite possible for animals to have bank accounts and check books—with certain banks —provided that some one is authorized to make out the checks for them. It is to discuss this matter that I have called you together, and I now propose that this measure be put to the vote."

Then followed a short discussion. Of course, the animals were all in favor of the Doctor's idea, with one or two exceptions; and the only active objections came from a tent-rigger and the clerk that sold the tickets at the circus box office. They said that they couldn't see what animals wanted money for, and they opposed the idea.

However, the other three human members present—the Doctor, Matthew and Fred—were in favor, and the motion was carried, on the human side, three votes to two.

In talking over this matter that same night at their after-the-show supper in the town house, the Doctor further explained his reasons for proposing such an idea.

"You see," said he, pouring himself out a second cup of cocoa, "I am not only thinking of providing for the animals in their old age, when perhaps I shall be no longer here for them to come to. But I hope by this to improve the standing of animals in general. There is an old saying, 'Money talks,' and—"

" 'Monkey talks,' did you say?" asked Gub-Gub.

"No. *Money* talks," the Doctor repeated. "It is a horrible thing, money. But it is also horrible to be the only one who hasn't got any. One of my chief complaints against people has always been that they had no respect for animals. But many people have a great respect for money. Animals with bank accounts of their own will be in a position to insist upon respect. If any one does anything unjust or unfair to them they will be able to hire lawyers of their own and prosecute the offenders in the usual manner."

"But how about their making out checks, Doctor?" asked Too-Too the accountant.

"I have already gone into that matter," said John Dolittle, "with lawyers and several bank managers. Most of them thought I was crazy and wouldn't listen to me at all. But two banks, the Eastminster and Chelsea, and the Middlesex Joint Stock Bank, agreed that if some person were given what is called power of attorney—and did the writing of the checks, the depositing and the drawing—those banks would have no objection to putting accounts in their books under the names of animals. Of course, about the lawyer part of it, if any prosecution should be necessary it remains to be seen what the courts will do. It would be something entirely new—and very interesting. I rather hope we do have a lawsuit come along soon so that we can see how it turns out. But in any event it will improve the standing of animals in the community, this having bank accounts of their own and lawyers of their own."

"He used to bother Too-Too four or five times a day"

Shortly after this arrangements were made for the animals' sharing in the profits of the Dolittle Cooperative Circus, and bank accounts were actually opened in the names of the animal members of the staff.

Gub-Gub, when he was handed a check book of his own, was highly delighted. He used to bother Too-Too four and five times a day to know how much money he had in the Eastminster and Chelsea. And he boasted to every pig he met that they had better be careful how they talked to him or he would write a letter to his lawyer and have him hauled into court.

Jip had an account opened at the Middlesex Joint Stock—and a very substantial balance he had when the Doctor had placed the money from Monsieur Poulain to the dog's credit.

Dab-Dab said she had no preference where she banked. At first she hadn't approved of the idea at all. But on consideration she decided it was an excellent thing, because she could thus keep part of the money safe from the Doctor's spending—to be used later, if need be, to get him back to Puddleby and a peaceful life.

THE SIXTH CHAPTER

Ways of Spending Money

IT was the Doctor's intention, once he had banked the money in the animals' names, to allow them to do what they liked with it, to give them complete control. To tell the truth, he was vastly interested to see what they would do with it. It was a new experiment.

It is most likely if the animals had been left to work out, each one by himself, how they should spend their new riches that there would have been a good deal of squandering. But although the Doctor did not influence them at all in the matter, they did influence one another quite a good deal.

On the evening of the day when they first came into their fortunes the Doctor was kept late at the theater talking with the managers; and the animal household sat down to supper in the kitchen of the town house without him.

"What are you going to do with *your* money, Jip?" asked the white mouse.

"I'm not quite sure yet," said Jip. "I've often had a notion that I'd like to set up a dogs' soup kitchen in the East End—a sort of free hotel for dogs. You've no idea what a lot of them are starving around the streets. A place where the waifs and strays could come and get a bone or a square meal and perhaps a bed for the night—and no questions asked—would be a good thing. I spoke of it once to the Doctor. And he said he would see what could be done about it."

"Oh, no you don't!" snapped Dab-Dab. "No more homes

for broken-down horses or stray dogs, thank you! I know what that means. You remember the Retired Cab-Horses' Association? That's the very thing we want to keep the Doctor away from. I'm going to leave my money where it is, in

"The animal household sat down to supper"

the bank. John Dolittle may be rich now—if what Too-Too says is true, he's as rich as the Lord Mayor of London—but nobody can get through money the way he can once he gets started. The day will come when he's poor again, never fear.

Then, if we've still got the money that he put in the bank for us, we'll be able to help him. He can say all he likes about our having earned it. But you know very well if it wasn't for him we wouldn't have a penny."

"With my money," said Gub-Gub, blowing out his chest, "I'm going to set myself up in a business as a greengrocer."

"Oh, for heaven's sake, listen to that!" groaned Dab-Dab, rolling her eyes.

"Well, why not?" said Gub-Gub. "I'm rich enough now to buy all the cabbages in England. I'm one of the wealthiest animals in London."

"You're the stupidest pig in the world," snorted the housekeeper. "And you'd probably eat all your vegetables instead of selling them. If you must go into the greengrocery business, for pity's sake wait till we've got the family back to Puddleby."

And so, in spite of the fact that they were the first animals to obtain the independence that comes with money, in the end, after they had talked things over among themselves, they did not show any disposition toward reckless or foolish spending. What new comforts they did provide for themselves were all of a reasonable kind. And the Doctor felt that this was a great triumph for his theory that animals, if treated properly, could behave quite as sensibly as people could.

But, of course, it would be unnatural to expect that the animals would not wish to buy something with their new wealth. And each in his own way (after he had questioned Too-Too about his balance) did a little shopping—just to celebrate, as Gub-Gub put it.

The white mouse's first purchase was a selection of foreign cheeses. He went with the Doctor to a very expensive West End grocery and bought a quarter of a pound of every kind of cheese that had ever been invented. The Doctor had never heard the names of half of them; and even to Gub-Gub, the great expert in foods, several were quite unknown. However, the pig made a note of their names and a description of their taste to be put in his new book, "The Encyclopedia of Food" (ten volumes), which he said he was now writing.

Gub-Gub's own investment was a selection of hot-house vegetables and fruits. He bought some of everything that was out of season, from artichokes to grapes. A lady customer who happened to be in the shop was very shocked at the idea of these delicacies being bought for a pig (Gub-Gub sampled most of them right away). And being a busybody, she remonstrated with the Doctor for giving such dainties to him.

"Tell her I've bought them with my own money," said Gub-Gub, sauntering out on to the street with his nose in the air and his mouth full of asparagus.

Another of Gub-Gub's new luxuries was having his trotters shined every day. An arrangement was made with a shoeblack, a small boy who lived not very far from the Doctor's town house. And every morning Gub-Gub stood very elegantly on the steps while his hoofs were polished till he could see his face in them. Dab-Dab was furious at this. She called it a vain, nonsensical waste of money. But the Doctor said it was not a very extravagant matter (it only cost a penny a day); and, after all, what was the use of Gub-Gub's having money of his own if he wasn't allowed to spend at least a little of it on frivolities?

"But why should he?" said Dab-Dab, her feathers bristling with indignation. "It doesn't do him any good to have his silly feet polished, and it certainly doesn't do us any good."

"What an idea!" said Gub-Gub. "Why, I'm the best dressed pig in Town."

"Best *dressed!*" snorted Dab-Dab, "when you don't wear any clothes."

"Well, I'm the best groomed anyhow," said Gub-Gub. "And

it's right that I should be. I have my reputation to keep up. Everywhere I go children point at me and say, 'Look, there's Gub-Gub, the famous comedian!' "

"The famous nincompoop, more likely," muttered Dab-

"Gub-Gub gets his trotters shined"

Dab, turning back to her cooking. "I suppose I can work my-self to death over the kitchen sink while that overfed booby goes mincing round the streets, cutting a dash."

"Well, but Dab-Dab," said the Doctor, "you don't *have* to do the housekeeping, you know—not now any more—if you would sooner not. We can afford to engage a woman housekeeper or a butler."

"No," said Dab-Dab, "I'm not complaining about that. Nobody else could take care of you the way I can. This is my job. I wouldn't want to let any one else take it over. The thing I object to is that stupid pig standing on the front steps every morning so the people passing on the street will see him having his shoes shined. He doesn't really enjoy being clean. He just thinks it's smart."

But some of the animals' first use of money was not that of buying in shops. Dab-Dab finally decided that she *would* hire a scullery maid to help her with the washing up. The housekeeper insisted that the new servant's wages should be paid for out of her bank account and not the Doctor's.

"After all, she is my assistant, and I ought to pay her," said she. "The Doctor's money will be needed for enough things without scullery maids' wages. Of course, she'll have to sleep out; I haven't room for her in the house unless I put her in with the pelicans. And they might object to that—a scullery maid—as soon as animals go on the stage they start putting on airs. Well, as the Doctor says, we shall see."

The scullery maid was engaged, and after she had been instructed in her duties by Housekeeper Dab-Dab (the Doctor acting as interpreter, of course) she fitted into the strange household extremely well. The effect on her of becoming a part of this theatrical family was that she became crazy to go on the stage herself. She wanted to sing in opera. Dab-Dab said she did all right for singing over the kitchen sink, but for opera she was hopeless. Nevertheless, she pestered the Doctor whenever she saw him to get her on to the operatic stage; until finally he arranged never to get home until after she had left, and in the mornings he would get Dab-Dab to inform him whereabouts in the house she was, so that he could escape from his room (where he took his breakfast in bed) and out into the street without meeting her in the hall or on the stairs.

Swizzle's first use of his bank account was rather peculiar and opened up still another new possibility for moneyed

"A smart nursemaid pushing a perambulator with five puppies in it"

animals. He discovered, while in the city (through some of the sick dogs who visited the Doctor's surgery on Greenheath), that he had a sister living in a suburb on the south bank of the Thames. Swizzle went to visit her when he learned of her whereabouts and took the Doctor along for a walk. He found that his sister had been married (it was seven years since he had seen her last) and now had a family of five very jolly little puppies. Two of them, however, were somewhat ailing, and Swizzle's sister (her name was Maggie) was very glad that the Doctor had accompanied her brother on the visit because she could now get the benefit of his professional advice.

"Well," said John Dolittle, "there's nothing seriously wrong with any of your children, Maggie. What they need is fresh air, all of them. London air isn't much to boast of at best. But, you see, this shed here where you have your kennel is very close and airless. You must get the puppies out more."

"But how can I, Doctor?" said Maggie. "They're scarcely able to walk yet. And even if they were, I'd never dare to take them on the streets till they have more sense—for fear they'd get run over."

"Humph!" said John Dolittle thoughtfully. "Yes, that's so."

"I tell you what, Doctor," said Swizzle suddenly. "Let's hire a nursemaid for my sister's family. Now I've got money of my own I can pay for one."

And that was what was finally done. A day or so later the citizens of London were provided with still another Dolittle surprise. A smart nursemaid was seen walking through the streets pushing a perambulator with five puppies in it, each wearing a warm knitted coat of white wool. Sometimes they were accompanied by their proud uncle, Swizzle, the clown dog from the Dolittle Circus.

The nursemaid was discharged, however, after her second week of employment and another engaged in her place. The puppies had complained that she would clean their ears in public, and they demanded a nursemaid who would be more considerate of their dignity.

THE SEVENTH CHAPTER

The Opening of the Animals' Bank

ONE more thing that came about through the idea of the Doctor's that animals should have money of their own was "The Animals' Bank." This institution did not enjoy a very long life but it deserves to be recorded in the memoirs of Doctor Dolittle as something well worthy of notice—as a remarkable event in the social history of animal life.

Jip it was who, always interested in the welfare of his fellow dogs of less fortunate circumstances than his own, first got the Doctor thinking over the plan. He had been speaking of his own experience (a story which you have already heard) when he first tried his hand at making money for the lame pavement artist.

"You know, Doctor," said he, "if I could earn money with that second-hand bone shop of mine, I don't see why other dogs couldn't do the same thing. But one of the reasons they never try it is that they'd have no means of keeping their money even if they made any. That's why I say, why not set up a regular Animals' Bank? Once it became established that animals had a right to hold property the same as people—well, there's no knowing how far the idea might go. You could have animals in all sorts of professions, earning salaries in jobs, making profits in private businesses and everything. But the first thing you need is an Animals' Bank."

"I see your point, Jip," said the Doctor. "I've always felt that horses, for example, had a perfect right to charge for their services in pulling carts—so much a load, you know. And there is no reason why watch-dogs should not receive the same pay as watch-*men*. They do the same work, and usually do it better. Well, what would you suggest that it be called, the 'Working Beasts' Savings Bank,' the 'Cat and Dog Trust Company,' eh?"

"No, I wouldn't have it the Cat-and-Dog anything," said Jip. "That sounds like a fight to start with. Besides, cats wouldn't go into it. They have no use for money. They'd never earn any. Cats are not public-spirited. They are naturally lazy. All they want is a soft place in the sun or a fire to sleep by. No, I'd just make it a general Animals' Bank, a place where any kind of creature, who wanted to, could bring his money—or goods to be turned into money—and know that it would be looked after and that he wouldn't be cheated."

"Yes, I think you're right," said the Doctor. "Well, we must see what can be done. It will be difficult probably to get a banking staff who will be willing to take over the duties of such an institution. Bank clerks are very particular, you know. The idea of cashing checks for horses and dogs may not appeal to them as in keeping with their dignity. And then it will likely be quite a job to get the financiers interested. You see, the animals' savings will have to be invested for them by the bank—otherwise they will get no interest and the bank won't pay. However, we shall see. I'll begin by going to that wealthy naturalist who lent me the pelicans and flamingoes from his private park. He has banking interests all over the city. He ought to be interested in the idea."

In what spare time he could give to it John Dolittle now set about the establishment of the Animals' Bank. He found the wealthy naturalist anxious to do anything he could to help.

"The advertisement alone, Doctor Dolittle," said he, after

" 'Yes, I think you're right,' said the Doctor"

the scheme had been explained to him, "will be well worth while for the furtherance of humanitarian treatment of animals. The idea sounds like a pretty wild one. But even if it prove impossible to keep up, it will be a good thing to have got it started. It will likely lead to other efforts from other quarters which may be more successful."

Thereupon he got in touch with some of his banking associates and persuaded them to give the plan a trial. A building in a good part of the city was rented and strong safes were set into the walls.

"THE ANIMALS' BANK, Incorporated," was placed above the door on a large signboard and in white enamel letters on the windows. Clerks and cashiers were engaged and desks and counters and money drawers were built for them. Ledgers and big account books were printed and bound with the name of the bank in gold on the backs—also check books were got ready in large numbers, especially made of strong rag paper so they would be able to stand the wear and tear of animals' use.

In addition to this, booklets, called prospectuses, were printed and sent out by mail in hundreds to tell the public that the bank had been organized, what it was for, and all about it. Notices were also printed in the newspapers that the bank would be opened to customers on a certain day.

This caused quite a little attention, criticism and comment. Jokes were made about the institution in all the funny papers, and editorials were written in a frivolous spirit in many of the biggest London dailies. But the Staff of the Animal Bank, backed by the wealthy naturalist, took no notice of this ridicule and kept right on with their preparations.

One of the results of the newspaper discussion was that the public became very curious to see how this extraordinary institution would succeed and how the difficulties of banking for animals would be overcome. To make sure that the affair should not be a failure, the Doctor had made careful plans. All his own animals had been given their share of the circus and opera profits the night before. And in spite of some of them already having bank accounts in other, older banks in London, they lined up before the doors were open,

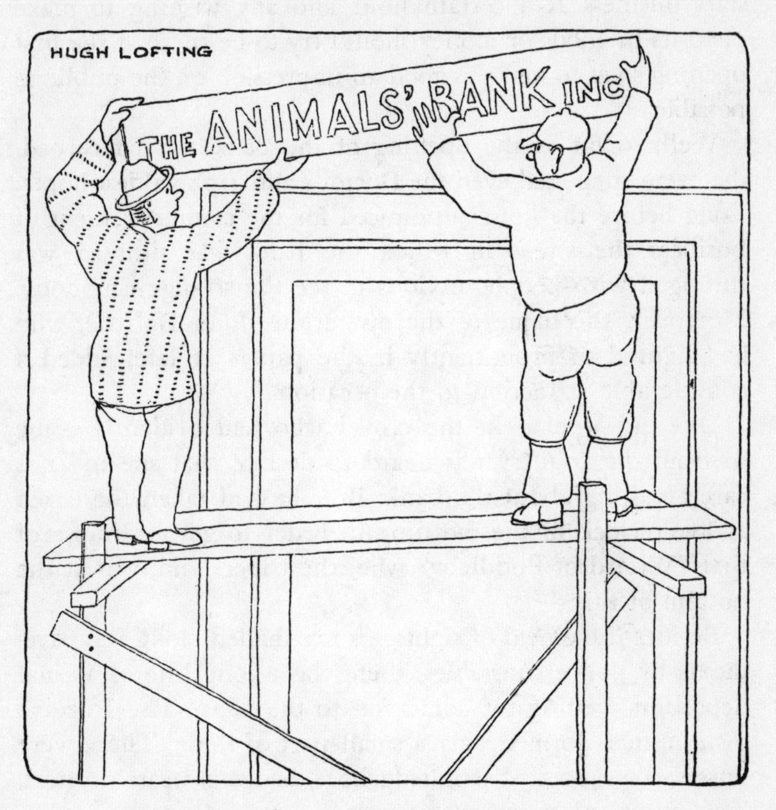

"A large signboard was placed over the door"

with their money for deposit, to show that the Animal Bank was an institution in which they trusted.

In addition to his own animals the Doctor had provided all the blackbirds with money to deposit and all the other birds of the opera cast, as well as a tremendous number of animals from London and the suburbs. Further, he had sent out word to all wild creatures in the country round about (by Too-Too and the white mouse) that the bank would start business at a certain hour and any wishing to make deposits in goods or money should try to be there at the first opening so as to make as good an impression on the public as possible.

Well, so far as the opening of the bank was concerned, the event surpassed even the Doctor's and Jip's wildest hopes. Long before the hour announced for the commencement of business the street in which the bank was situated was thronged with people anxious to see the strange ceremony. Moreover, the name of the mysterious John Dolittle, who had figured so prominently in the papers of late, added a considerable attraction to the occasion.

One old woman in the crowd who had evidently come in from the country was heard to declare that she did not care anything about the bank. But she had taken the coach at five o'clock in the morning in order to get a glimpse of that "Wizard of Puddleby" who, the papers said, was at the bottom of this.

Besides the crowd of sightseers marshalled along the pavements by police constables, there was a long line of animal depositors waiting for admission to the bank. The Doctor's own animals formed only a small part of these. There were horses, cows, dogs, sheep, hedgehogs, badgers, weasels, otters, hens, geese and many other kinds.

The Doctor remarked to Matthew and the wealthy naturalist that already their plan was working—even before the bank opened—because none of the animals were being molested by the people. Ordinarily, he said, the boys and idlers on the street would have teased or tried to catch some of these wild creatures who were strangers to city eyes. But the fact that they were here to deposit money in the bank like ordinary citizens already made the people treat them with respect. And although the crowd cracked many a joke as some timid hare or cheeky jackdaw came to

"He received a check book from the manager himself"

join the waiting line, nobody attempted to annoy them or interfere with their business rights.

Finally, as a neighboring church bell struck the hour of nine, a little wave of excitement passed through the crowd.

The big doors of the bank were seen to be swinging open. The throng surged forward to get a glimpse through the windows of what would happen within. Policemen had to form a line to keep them back. Reporters from the newspapers clamored for permission to get nearer so that they could make pictures of the first animal going into the Animals' Bank.

It was one of the rules of the bank that every animal wishing to deposit money must be accompanied by a person having power of attorney. And as each depositor came up to the door the Doctor or Matthew went in with him and gave his name at the cashier's desk.

There were several thrills during the course of the morning's work which made the crowd feel that the show had been worth waiting for. One was when a badger came forward with over a hundred ancient gold coins (old buried treasure) which he had dug up while making his hole in the side of a hill. The gold was weighed and its value deposited to the badger's credit. Another stir was caused when a closed van drove up to the door through the dense crowds and a live African lion got out, accompanied by Fred (it was the Dolittle Circus lion). With great dignity the King of Beasts strode into the Animals' Bank, deposited ten pounds and received a special gilt-edged check book from the bank manager himself—who also made a short speech. It was the star turn of the day's business and was reported (with pictures) in all the evening papers throughout the city.

THE EIGHTH CHAPTER

News of Puddleby

BUT of course for the upkeep of a regular bank a tremendous lot of constant business is a necessary thing. And although the number of depositors appearing on the opening day was very considerable, the amount of money they put in did not, even when it was all added together, make a big sum.

Further, many of the animal customers were so delighted to have check books of their own that they immediately began writing out checks to see how it worked; and with those whose balance was only a small amount to begin with this soon brought them back to where they were before.

And then after the excitement of the opening day was over the number of daily customers doing business at the bank fell off noticeably. It was a difficult thing for out-of-town depositors to get into the city. The general manager of the Animals' Bank, Inc., told the Doctor at the end of the first week that if the institution were to pay the ordinary interest on the customers' money, a much larger amount of business would be necessary.

However, the wealthy naturalist was determined to keep the institution open for the animals and the Doctor a fortnight at least. And he promised to make good any loss that the bank might suffer and to see that the clerks and staff were paid their salaries—even if the bank made no profits at all.

Dab-Dab was very anxious over the Doctor's part in the

business. She repeatedly told Jip and Too-Too that she wouldn't feel safe and happy until she got him back to Puddleby.

" 'Of course,' said Dab-Dab, 'I can't make him sensible in money matters' "

"Of course even then," she said, "I can't make him sensible in money matters. But he will have less temptation for spending. Here in London, with so much going on, you can't tell from day to day what he may do next. It was only by the greatest stroke of luck that I heard he was going to the rescue of the Animals' Bank with his entire fortune when he heard they were near to failing. He said he would be the one human depositor there and he did not want his naturalist friend to lose any more money when he had already been so helpful. It took me over an hour to dissuade him from the crazy notion."

But while the comic papers had a good deal of fun over the closing of the Animals' Bank, neither the animals themselves nor any one who had helped with the starting of it felt that it had been by any means a failure. Matthew Mugg, who was always on the lookout for chances for advertisement, kept the last week of the bank's career full of interesting incidents. Many societies whose purpose was the prevention of cruelty to animals and similar aims were invited to inspect the bank and meet the Doctor. These occasions were all reported in the newspapers and served to keep the name of the Canary Opera and Dolittle Circus before the public.

Matthew also announced at the Doctor's orders that the Circus and Opera would probably go out of business before very long. And this the Cats'-Meat-Man did in such a way that the public, feeling this was their last chance to see these wonderful shows, thronged to the Regent's Theater and the Greenheath circus grounds in larger numbers than ever.

It was shortly after the events just described that Gub-Gub started still another new thing and nearly drove Dab-Dab crazy. One evening after supper the now wealthy pig-comedian suddenly said:

"Doctor, I have an idea."

"One of your best?" asked Jip, who was warming himself by the fire.

"Yes. One of my very best," said Gub-Gub, carefully choosing an apple to take away from the table for eating in bed.

"Listen: we have had the Canary Opera which was a more or less serious affair—very successful—but now about to be closed. Why not let us have a Food Opera—a comic opera?"

A chorus of laughs came from the household. But Dab-

"Carefully choosing an apple for eating in bed"

Dab seemed in no mood for nonsense. She approached the great comedian and pushed her bill close up to his nose.

"If you dare," said she, "to start any new foolishness to keep the Doctor away from Puddleby any longer you won't get a thing more to eat while I'm in charge of the house-keeping."

"Well, but wait a minute," said the Doctor, laughing. "There's no harm in hearing what Gub-Gub has to suggest, even if we don't produce his Food Opera. What kind of a show did you mean it to be, Gub-Gub?"

"Quite light," said the pig. "In fact comic throughout, with the exception of one poetic little scene where food-fairies dance round the spring lettuces in the moonlight."

"But I don't see how you could write an opera entirely around food," said the Doctor. "Describe some of it for us."

"Well," said Gub-Gub, "for instance, there was one number called the Knife and Fork Quartette. You know what a lot of noise some people make with their knives and forks at table—and then they always lay them together on the plate, to show that the concert is over. Well, I thought if one had four people and trained them to eat in time, all together—it's a pleasant noise after all, the noise of a knife and fork—you could work out a very interesting quartette."

"I see," said the Doctor. "And what else?"

"Then there was a soup chorus which I had in mind," Gub-Gub went on. "I've noticed everybody drinks soup in a different key—some very musically. The treble parts could be drunk by small dogs like Toby here. And the bass parts by—"

"Pigs," put in Jip.

"How unspeakably vulgar!" muttered Dab-Dab. "For heaven's sake, Doctor, send that pig to bed before I forget I'm a lady!"

"Any dances in your show, Gubby?" asked Swizzle—"Nothing but guzzling?"

"Certainly there are dances," said Gub-Gub. "There is a napkin ballet in the first act, and a very grand Waiters'

March in the second—besides the Caper Sauce Caper at the finale. Oh, there are lots of good turns in the Food Opera. One of the arias is called, *Songs My Kettle Used to Sing.* Another is entitled, *Poor Little Broken Pie Crust*—very light —sung by a comic character called Popper Popover. Then there's a love ballad which begins, *Meet Me in the Moonlight by the Garbage Heap.*"

"Your opera sounds very indigestible to me," said Jip.

"Well," Gub-Gub explained, "it isn't meant to be grand opera. It's light opera—not a heavy meal. Mine would be the kind of show they call in French *opera bouffe.*"

"Opera beef, I should call it," growled Jip.

"You know, Dab-Dab," said the Doctor, "there's something in it. Gub-Gub has ideas."

"I always thought I could write," muttered the pig beneath his breath.

"And, you see," the Doctor went on, "now that Gub-Gub has a name, the fact that the opera was writtten by him would have a good deal of weight with the public."

"Lord deliver us!" cried Dab-Dab. "Haven't we had enough of the stage for a while? Haven't we had enough success—made enough money? The London playgoer isn't going to come to crazy animal shows forever. To me this looks like a fine thing for you to lose all your money on, Doctor. Why can't you let well enough alone? You want to go back to Puddleby—we all want to go back. Very well, then, let's go while we've still got money enough. This pig's vanity is running away with him. And you're letting it run away with you too."

"We could have a splendid orchestra," Gub-Gub went on thoughtfully, "with all the kitchen utensils—dish covers and tinkling wineglasses."

Happily for poor Dab-Dab, who was almost in tears at

the prospect of the Doctor's launching into a new theatrical enterprise on the eve of his leaving the show business altogether, Cheapside, the London sparrow, suddenly turned up while the discussion was still going on.

"Mr. and Mrs. Cheapside"

That excellent chorus-master had a week before taken a run down to Puddleby with his wife Becky. The pelicans and flamingoes had, of course, become perfect in their parts after more than a hundred performances, and Cheapside had decided that a few days' change in the country would do both him and his wife good. Now, immediately the two sparrows appeared, the entire party, including the Doctor himself, forgot all about the Food Opera and clamored for news.

"Well," said Cheapside in answer to the general chorus of questions, "the country looks lovely—just lovely—though, of course, there ain't no flowers to be seen yet 'cept a few sprigs of hawthorn here and there. But the trees is budding green all over, and Spring is started all right. The garden?—Well, the less said about that the better. There was a half-dozen crocuses pushing themselves up through the lawn, but it's 'ard work for the pore things. Break anybody's 'eart it would, to push 'is 'ead up through *that* lawn and see the mess. Hay —old dead hay from last year, a foot thick, is all that's left of your lawn, John Dolittle. If you don't go back soon and put that garden straight you won't be able to find the house for weeds."

Dab-Dab, saddened though she was at Cheapside's description of the beloved home, drew a good deal of satisfaction from noticing that the Doctor too seemed deeply impressed by Cheapside's words.

"How about the apple trees, Cheapside?" asked John Dolittle presently, after a moment's silence.

"Pretty far gone, Doc," said the sparrow. "A good gardener might be able to save 'em, if he got after 'em right away. They ain't been pruned in so long they look more like old men with beards on than apple trees. But the blackbirds and thrushes are building in 'em just the same."

"Humph!" muttered the Doctor. "And the old lame horse, he's all right of course?"

"Yes," said Cheapside. "He's all right, in a way, but he gave me a message for you. Told me to be very particular how I gave it, too, because he didn't want you to be offended.

It seems he's lonely. Gets all the food he wants, of course—goes out on the back lawn whenever he wants grass, and since you wrote to the hay and feed merchant in Puddleby he's had oats and all sorts of grain, as much as he could eat. 'But,'

" 'Well, now,' said Dab-Dab, bustling forward"

'e says to me, 'e says, 'tell the Doctor I ain't complaining, but I've heard about this Retired Cab and Wagon Horses' Association what he started on a farm of their own up Kettleby way. An old friend of mine, a fire horse, what used to pull the fire engine in Great Culmington, has joined. And, well, I ain't complaining, but it's awful lonely here all by myself with nobody to talk to. And if the Doctor's going to be away much longer I'd like to be sent up to the Association's farm. I'm getting kind of tired of talkin' to meself.' "

"Well, now," said Dab-Dab, bustling forward as Cheapside ended, "there you are, Doctor! We're going to lose one of our oldest friends if you don't get back—besides having the finest garden in the West Country go to ruin and the house fall to pieces. I tell you if you don't go back soon you never will be able to make that place into what it was. It will just get beyond repair."

"Yes, I suppose you're right," said the Doctor. "And goodness knows, I'm dying to see the dear old place again. Well, Gub-Gub's opera will have to wait. Maybe even when we've gone out of the show business we may try it out some time in Puddleby, a private performance, perhaps to celebrate the author's birthday. Now, listen, Cheapside, I wish you'd fly down to Puddleby again to-morrow and tell the old lame horse that I'm coming back just the earliest moment I can. We have some clearing up matters to attend to here before we can leave. I must see that the menagerie animals are properly provided for, of course. But just as soon as ever we can get away we're all coming back to Puddleby."

THE NINTH CHAPTER

The Legend of the Jobberjee Ghosts

CLOSING up a circus is no small matter. There are a terrible lot of things to be thought of and provided for. Poor Dab-Dab grew more and more anxious, as the preparations proceeded, lest at the last minute something should crop up to alter or put off the Doctor's intention of retiring from the show business.

But it did really look as though this time John Dolittle himself were determined that nothing should interfere.

Still it was more than any one could hope for, of course, that the Doctor could get through such an enormous undertaking without extravagance (as Dab-Dab called it) of some kind. And the most costly item on the list of his expenditures was caused by the wild animals of the menagerie. Of these now remaining there were only three, the elephant, the leopard and the lion—all Africans. You should have heard the rejoicing in the menagerie when the Doctor announced to them that he intended sending them back to freedom and their native land! The leopard sprang around his cage yelping for joy; the lion roared so that folks on Greenheath thought that the end of the world was at hand; and the elephant, trumpeting at the top of his voice, executed his stage dance with such abandoned enthusiasm that he nearly wrecked his stable. Gub-Gub said it reminded him of the night when the menagerie animals helped Sophie to escape by kicking up a prearranged rumpus.

But it was not merely the sending of the animals back to Africa that was so expensive—though, to be sure, even that would have cost a lot of money. But John Dolittle was determined that these performers, in return for their faithful services in his circus, should travel in luxury. He very strongly disapproved of the usual manner in which animals were made to travel—he said there wasn't one of them that was given decent quarters, from chickens to elephants. And his animals were going to have a ship of their own!

Poor Dab-Dab! She did not know exactly how much a ship big enough to carry an elephant, a lion and a leopard would cost; but she knew it would be expensive. She threw up her wings in horror.

"Why on earth can't they go by the ordinary liners that ply between London and Africa?" she squawked.

"Because, Dab-Dab," said the Doctor, "no ordinary ship can possibly give room enough to animals of that size if they carry other freight or passengers. And, besides, these ordinary packet boats and liners only go to the big ports, where my animals would probably be captured or shot as soon as they were landed. I want them, after their years of service, to be accommodated properly on the voyage and set ashore exactly where they want to land."

And so in spite of all protest from his housekeeper the Doctor advertised in the London papers for a ship. He received dozens of answers by mail. But in choosing the craft, the captain and crew he was very particular. Many of those who answered the advertisement refused, when they found out what freight they were to carry and what were the conditions under which they were to sail, to take on the job at all. The Doctor insisted that the animals were to be given the run of the ship and that they would only be confined below decks when the weather was stormy. Their comfort in everything was to be the chief consideration throughout.

But a few captains when they learned that Fred, the menagerie keeper, was going to accompany the animals (the Doctor was sending him to make sure that his friends were

"The Elephant makes the acquaintance of the captain"

properly treated) were willing to consider the proposition. Then came the task of finding out whether the animals themselves approved of the ship and the crew.

This John Dolittle did by first visiting several of the craft himself, and after he had picked out two or three of the best of them, he made arrangements to have the elephant, the lion and the leopard go down to the docks in their circus traveling wagons to look over their quarters and make the acquaintance of the captain and the sailors who were to take them.

The second ship that the animals looked over they declared themselves very pleased with. It was the one that the Doctor himself thought the best, too. The captain was a good-hearted old salt, and after he had been introduced to the animals and been shown that they had no intention of molesting any one on board, he said that he would personally see to it that they were treated like first-class passengers and given the best of everything.

The Doctor arranged with Fred and the animals a system of signals whereby they could make their wants known to him and he would convey them to the captain and the crew.

It was a good thing that the Doctor had made so much money with the Opera and Circus. Because although he had expected that this chartering of a special vessel would be a very expensive matter, the actual sum required turned out to be a good deal more even than he had anticipated. For one thing, for carrying such large animals as elephants and lions almost the whole ship had to be refitted and altered. Carpenters were busy on her for weeks before she sailed.

A most luxurious stall was made for the elephant amidships, with specially strong padded sides, in order that he would not be banged about if the weather got rough. In the fo'castle an enormous shower bath was put in for him, so he could keep cool in the hottest weather. The quarters for the lion and the leopard were similarly provided with the most up-to-date kind of traveling luxuries.

Then the provisioning for such a long voyage was another considerable matter, especially as the Doctor wanted

"The Doctor arranged between Fred and the animals a system of signals"

the animals to have every delicacy they wished for on this last journey they would make in human care.

However, the Dolittle bank account was now, for a while anyhow, big enough to stand even this with ease. And although Dab-Dab kept pestering Too-Too the accountant for statements, the Doctor and he were able to show her that there would still be lots of money left.

The voyage, as Fred described it to the Doctor on his return, was quite pleasant and successful. Good weather favored the travelers nearly all the way, with the exception of a few rough days crossing the Bay of Biscay, when the elephant kept to his bunk in his luxurious cabin and the lion lolled on a specially constructed couch in the saloon, sipping chicken broth. For the rest of the voyage the animals were able to stay on deck all day long. Their appetites from the sea air grew enormous, and the provisions were only just enough to last them comfortably, although a very large stock had been laid in.

As to their exact destination, even the animals themselves were unable to instruct the Doctor very definitely, although he had done his best to get from their descriptions of their homes some idea of what point on the African coast the captain should make for. Because, of course, their knowledge of geography, outside of their own particular country, was very small. So the best that could be done was to make landings at various parts of the coast of Africa and let the animals go ashore to see if they recognized it as their own home district. The Doctor knew that the leopard probably had been captured somewhere in West Africa, and he told the captain to stop in at Sierra Leone first and see if that was not his home. And sure enough, the leopard at the first port of call sped off into the jungle with a yowl of delight and was not heard from again.

The finding of the homes of the elephant and the lion was not, however, so easy. The lion had described his home as a mountainous part of Africa, where the woods were broken up with open stretches of park land and many nice small streams of good water. The elephant said that the thing he best remembered about his particular section was that the grass grew very high (up to his shoulders and higher), and that generally it was flat or gently rolling. From this the Doctor concluded that both of them came from somewhere between the Zambesi and the Djuba rivers. But, as this was a very long stretch of coast, many landings had to be made before the animals were sure that they had reached the country of their birth.

Putting an elephant ashore when you have no proper wharf to moor your ship to is quite an undertaking. So a good deal of time was spent in making the various stops.

On his return Fred told the Doctor that during several of these halts, when they were engaged in hoisting the elephant's huge bulk over the side with a sling and a derrick, government officials and coast-guardsmen came up and wanted to know what they were doing. When it was explained that a wealthy circus owner had sent a special ship down to put his performing animals back on their native soil, the government officials said he must be very rich and very crazy and went on.

At each of these landings, of course, Fred went ashore with the animals to see if they had come to the right place. And finally, about a hundred miles south of the mouth of the Zambesi River, both the lion and the elephant sniffed the wind as soon as they landed and generally behaved in such a way that Fred decided they had come to the right place at last. But, he told the Doctor, he was somewhat surprised at this, because from their description, it would appear that the two had come from different parts of Africa. After climbing a small hill, where they could get a good view of the surrounding country, they seemed to be quite certain that this was the land they were seeking. They evidently wanted to show Fred, before they left him, that they were grateful for his kind treatment of them, and he gathered that they were also trying to give him some farewell message to carry to the Doctor. But his knowledge of their

"Standing on a tree stump waving his trunk"

language and signs was so poor that he could not get the meaning of what they were trying to say. At last, frisking and frolicking like kittens at play, they plunged down the hillside together and disappeared into the thick bush.

Fred said that he felt sure that they had only reached the lion's country, and not the elephant's though, of course, the two species of animals often inhabit the same areas. And he decided that the elephant, after his long friendship with the lion, had determined to stay with him for company.

And this may very likely account for the strange legends and tales that the Doctor later heard from natives of that part of Africa. Many hunters and trackers declared that they had often seen a lion and an elephant going through the jungle together as though they were inseparable companions. And that at night-time they frequently came down together to the salt licks, where many wild creatures were gathered, attracted by that particular delicacy. Here in the weird African moonlight they would give a strange performance, which seemed greatly to entertain the other denizens of the jungle—the elephant doing a curious dance, standing on a tree stump waving his trunk from side to side as though music were being played for him, while the lion gave jumping exhibitions, springing up on to the elephant's back and down the other side. This extraordinary team of jungle performers came to be called by the black men the "Jobberjee Ghosts."

The white sportsmen who came there after big game paid very little attention to these stories of the mysterious Jobberjees, considering them merely the wild imaginings of scared, superstitious natives. But the Doctor knew that they were probably quite true—that his circus animals, who in the Dolittle show ring ran their own performance without the aid of whips or human orders, had carried their tricks into the heart of the African jungle.

THE TENTH CHAPTER

The Dolittle Circus Folds Its Tents for the Last Time

DURING his stay in London and Greenheath the Doctor had, as I have told you, attended to a very large number of animal patients, who began and continued to come to him as soon as the news of his arrival had spread. These were, of course, mostly city animals, such as dogs and cats and rats and those birds who make their homes near the haunts of Man.

For the Doctor the dogs were the most interesting. Jip used to love to go off by himself, wandering around the back streets and the slums of London; and whenever he met a sickly-looking mongrel or a stray that seemed to have no home he would get into conversation with him and always ended by giving him the Doctor's address. Jip was thoroughly democratic and very charitable. And he believed in giving his fellow creatures whose fortunes had fallen low a helping paw.

This, of course, was very commendable, but it was not approved of by Dab-Dab at all. Every time the Doctor got back to his circus wagon at Greenheath he found it simply surrounded by the tramps and outcasts of dogdom. Some were sick or lame and needed medical attention. But the greater part of them had just come, hoping to get a glimpse of the great man, a square meal or a bone. But all brought with them a secret hope in their hearts that the Doctor would adopt them and take them into his famous household.

In spite of the work and the time it took up, the Doctor was always glad to see them. His only regret was that he could not take every one of them as part of his animal family —as Jip hoped he would. Jip was never jealous of his famous master's attention. It was his belief that the Doctor could not have too many dogs.

"After all," he used to say, "dogs *are* the most intelligent animals to keep. Why not have plenty of them?"

It was while the Doctor was still in the midst of his preparations to leave the circus life and return to Puddleby that Jip once more suggested his idea for a Dog's Free Food Kitchen in the East End of London. He wanted to spend his own money on it, he said. But the Doctor explained to him that the cost of keeping a free kitchen open continually would be much more than he expected.

"Why, Jip," said John Dolittle, "you would have to have at least one cook on duty all the time, and his wages alone, without counting anything for provisions, would run into a considerable amount of money per year!"

"Well, perhaps we could do without a cook, then," said Jip. "Many dogs would just as soon have their meat raw. Bones are always better uncooked."

"Humph!" said the Doctor. "Possibly, if you brought the idea down to just a bone distribution station and got a boy to run it for you, it would not be so expensive—that is, for a while, anyhow. But your bank account would not be big enough to keep it open forever, that's certain."

Jip was very disappointed. However, he was determined to be a philanthropist in a small way, even if he couldn't keep it up. And, after making inquiries through the Doctor as to the wages that boys got, he went over estimates and costs with Too-Too the accountant and found that he could run a Free Bone Counter for a week without beggaring himself.

So he set to work and hired a healthy strong boy. Then he rented a cellar in Whitechapel and opened an account with a butcher who was to supply the bones fresh every day. These business transactions, of course, he did through the

"So he hired a healthy strong boy"

Doctor, but he was very particular that all expenses came out of his own money.

Thus came into existence—even though a brief existence only—Jip's *East End Free Bone Counter for Indigent Dogs.* It was a great success while it lasted. Jip himself was in attendence at the cellar door almost continually. And night and day a constant procession of half-starved mongrels traipsed up and down the steps, carrying away bones with them, ranging in size from a lamb cutlet to a shoulder of mutton. Jip was careful to inspect all the dogs as they came in, to make sure that they were deserving cases. If the dogs wore smart collars or had well brushed coats he knew that they were owned by rich folks and were merely coming out of curiosity. These he chased away. But to the rag-tag and bob-tail of dog society he was most hospitable and kind, often waiting on the poorest of them himself.

When it became known at Jip's bone kitchen that the Dolittle household was shortly leaving London for Puddleby the Doctor was besieged with applications from dogs who wanted to come with him. What made poor Dab-Dab particularly angry was that Matthew Mugg, who had a great love for dogs of every kind, aided and abetted Jip in trying to get the Doctor to take every mongrel that applied.

"I'd like to know, Jip," said she, "where in heaven we would put all those disreputable creatures if the Doctor only took half of them. I should have thought you would have more sense. The Puddleby house is crowded, as it is. Yet for every mongrel that asks to be taken you put in a good word, and that stupid Matthew, too, just as though the house were as big as a hotel."

"Oh, well," said Jip, "the garden's large enough—even if the house is small. They could camp outside in kennels, or something."

"Yes," snorted Dab-Dab angrily, "and a fine garden it would be—with hundreds of half-breeds rooting around in it! A garden is for flowers, not for the riff-raff of London streets."

Well, finally the Doctor said that perhaps later, when they were properly settled back at Puddleby, he might take a piece of land and start a Home for Cross-Bred Dogs similar to the rest farm for retired cab and wagon horses. And with this Jip had to be content.

Getting the rest of his circus animals properly disposed of

"He was besieged with applications"

was not so hard for the Doctor as had been the case with the large menagerie animals. Through his rich naturalist friend who had lent him the pelicans and flamingoes he heard of a certain Mr. Wilson, a reliable person, who was shortly sailing for America. With him the Doctor made arrangements for the opossum (whom Blossom had styled "the famous Hurri-Gurri") and the six American blacksnakes which Fatima had called king cobras. These, while they were not exactly the easiest kinds of pets to travel with, could, after all, be made comfortable in properly constructed boxes which would be small enough to go as personal baggage.

Mr. Wilson was not a naturalist in any sense of the word. But he promised the Doctor as a special favor that he would see to it that the opossum and the snakes were properly cared for on the voyage.

John Dolittle did not realize that he was giving Mr. Wilson anything but a very simple commission to carry out. But, as a matter of fact, these creatures caused quite a little sensation of a harmless kind before they were returned to their native soil. On the voyage, while the opossum was being given an airing on the deck in charge of one of the stewards, he decided that the ship's mast was a new kind of tree. And suddenly he scaled up it at tremendous speed and began hanging from the ropes by his tail. The steward, who was not good at climbing, called him, but he refused to come down. A crowd of passengers gathered on the deck to watch the show. Finally a sailor was summoned by the captain and sent aloft to capture the opossum. But he was not easily caught. He liked the view from his new quarters in the rigging, and he meant to stay there for the rest of the voyage.

Seeing that the sailor was trying to take him prisoner, he kept swinging from mast to mast, running around the rigging as though he were on dry land. The sailor, good climber though he was, stood no earthly chance of catching him single-handed and he called down for more assistance. Then another deck-hand was sent up. But still the agile little animal kept out of their grasp.

Finally all the crew that could be spared were told off for the hunt. But even with six sailors trying to capture him the opossum continued to give his pursuers the slip and to enjoy the view.

Night came on and still Mr. Opossum was in the rigging and the six sailors had grown very tired of climbing about on ropes. All they had succeeded in doing was to give the passengers a very amusing show to watch. With the coming of darkness the hunt had to be given up.

However, during the night a cold wind sprang up and the opossum found that his lofty quarters were a trifle chilly. So he came down of his own accord. He was found by his steward, shivering behind a ventilator and put back in his box in Mr. Wilson's cabin—to which warm comfort he was very glad to return.

But Mr. Wilson's troubles were not over yet. When the ship docked at New York and his luggage was opened and examined by the customs officials the six blacksnakes got out and started wriggling all over the dock, delighted at a chance to stretch themselves after the confinement of the voyage. Everybody was scared to touch them. And a special man from the Zoo was sent for to get them back into captivity.

When he arrived and the snakes realized that they were being chased they, too, got scared and, diving among the passengers' baggage, they tried to hide behind trunks and handbags as best they could. One of them got into an old lady's valise which was opened for inspection and its owner went into a faint when she saw a four-foot blacksnake squirming around among her laces and shawls.

But finally they were all captured and Mr. Wilson commissioned the man from the Zoo to take them and the opossum into the country and set them free.

The Canary Opera had some time before this, of course,

given its last performance and disbanded. The thrushes, the wrens and the rest of the birds who had helped had taken wing back to their natural haunts. The Dolittle town house had been given up and charwomen were cleaning it, after

"A throng of youngsters came with an enormous bouquet"

its five months' duty as an operatic aviary. On Greenheath the Dolittle Circus enclosure looked very empty and deserted. Most of the tents and wagons and caravans had been sold at auction and hauled away. The mechanical merry-go-round and the Punch-and-Judy theater had been presented by the Doctor to a school for foundlings. John Dolittle's own wagon (the original one that Blossom had had made and painted specially for him) and the wagon occupied by Mr. and Mrs. Mugg were, with the Pushmi-Pullyu's stand, about all that was left of a once gay and elaborate circus.

Matthew said he just couldn't bear to look at the scene, it saddened him so. But his wife, Theodosia, said she didn't see what he had to complain of—even if the days of the Dolittle Circus were over—now that he was better off than he had ever been in his life.

Many hundreds of children living in the neighborhood of Greenheath (who had frequently visited the circus which the Doctor had designed particularly for young folk) were, even more than Matthew, saddened at the prospect of its closing up for good. On the day before the Doctor announced he would leave, a vast throng of youngsters came, with an enormous bouquet of flowers, to bid him good-by. And this, when he walked down the steps of his wagon and for the last time distributed the free peppermints to the children, was (he told Matthew afterward, when they were on the road for Puddleby) the only thing that made him feel sorry at leaving the life of the circus behind.

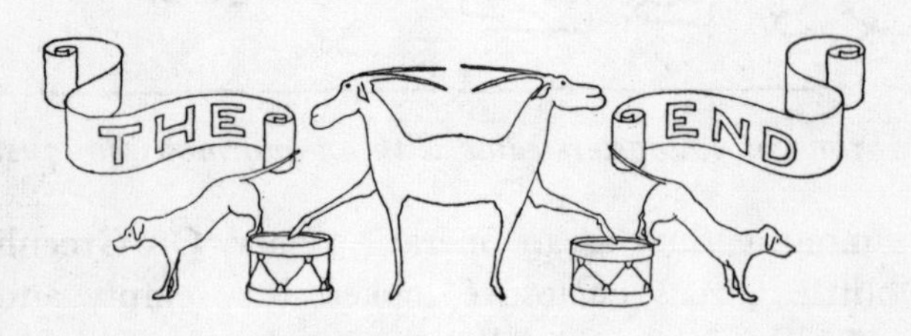